Hard X-Ray Imaging of Solar Flares

Michele Piana • A. Gordon Emslie
Anna Maria Massone • Brian R. Dennis

Hard X-Ray Imaging of Solar Flares

Michele Piana
The MIDA Group, Dipartimento
di Matematica
Università di Genova and CNR - SPIN
Genova
Genova, Italy

Anna Maria Massone
The MIDA Group, Dipartimento
di Matematica
Università di Genova and CNR - SPIN
Genova
Genova, Italy

A. Gordon Emslie
Department of Physics & Astronomy
Western Kentucky University
Bowling Green
KY, USA

Brian R. Dennis
Solar Physics Laboratory, Code 671
Heliophysics Science Division
NASA Goddard Space Flight Center
Greenbelt, MD, USA

ISBN 978-3-030-87279-3 ISBN 978-3-030-87277-9 (eBook)
https://doi.org/10.1007/978-3-030-87277-9

This Springer imprint is published by the registered company Springer Nature Switzerland AG
The registered company address is: Gewerbestrasse 11, 6330 Cham, Switzerland

Foreword

High-energy solar flare physics has had an incredibly exciting journey over the past decades mainly driven by new observations from NASA's Small Explorer mission *RHESSI*; this book guides researchers along the path of this journey to our current understanding.

By astronomical standards, the Sun is in front of our nose, making it possible to study not only its spectral features, but also to obtain detailed resolved images. With the recent launch of ESA's Solar Orbiter mission, we now have observations from different vantage points allowing us to stereoscopically measure solar flares. This makes our Sun an exciting astrophysical object, and despite the many decades long advances that have been achieved, there are still many discoveries to be made.

Hard X-ray observations provide a key diagnostic tool to study the energy release in solar flares that provide quantitative measurements that are fundamental to the physical processes at work. However, imaging at hard X-ray wavelengths is challenging and observatories generally use indirect imaging methods. Compared to observations at other wavelengths where focusing optics can be readily used, hard X-ray images need to be reconstructed before they can be implemented in solar flare studies at multi wavelengths. Indirect imaging provides an exciting opportunity to develop cutting-edge algorithms optimizing the reconstructed images and therefore the science output.

The indirect imaging approach however has proven to be challenging for users outside the core instrument teams, as it often appears to be a black box for the general user. In addition, scientific publications do not properly address the fundamentals of imaging reconstructions, or even less so the hidden details of how to best apply the different algorithms. This has been a challenge for our solar flare community and it has limited our growth, despite the recent exiting advances achieved. With this book, the gap between the expert users and the solar flare community will be bridged. I am very pleased to see this book being published, and it will become an essential companion for anyone working with the *STIX* instrument onboard ESA's Solar Orbiter mission. I sincerely thank the authors for putting in the effort to make it happen.

Principal Investigator, *RHESSI* and *STIX* Säm Krucker
FHNW
Windisch, Switzerland
October 2021

Preface

The idea for this book emerged over a number of years as the authors participated in various research projects related to analysis and interpretation of data from NASA's *RHESSI* Small Explorer mission. The data produced over the (unexpectedly long; February 2002–April 2018) operational lifetime of this mission inspired a large number of investigations related to the overarching science question that *RHESSI* was designed to address: the when, where, and how of electron acceleration in the stressed magnetic field environment of the active Sun.

As discussed in Chap. 1, a vital key to unlocking this science problem is the ability to produce high-quality images of the so-called "hard" X-rays produced by bremsstrahlung radiation from electrons that are accelerated over the course of a solar flare. To adequately address the science objectives, these images must cover the energy range from less than 10 keV up to more than 100 keV with keV energy resolution, spatial resolution in the arcsecond range, and a time cadence of order 1 s. The only practical way to do this within the technological and budgetary limitations of the *RHESSI* era was to eschew imaging via some form of focusing optics in favor of the indirect imaging technique described in some detail in Chap. 3. This technique involves the use of multiple "rotating modulation collimators," through which imaging information is encoded, not in the usual "pixel-by-pixel" fashion of direct imaging devices, but rather as a set of modulation "patterns" produced by rotating an instrument made up of multiple bi-grid X-ray collimators, with each one placed in front of a detector that has no inherent spatial resolution. It can be shown that each such pattern corresponds to a different two-dimensional spatial Fourier component of the source and can be identified through the appearance of a particular signature in the time profile of the measured count rate in the corresponding detector. The image is reconstructed from this information using one of many computational techniques that have been developed, in many cases expressly for this purpose.

It is fair to say that this way of thinking about imaging was not one with which the solar physics community was entirely comfortable, at least initially. Radio astronomers, working with large interferometer arrays, had employed Fourier imaging techniques for many years, primarily for observations of astronomical sources other than the Sun. They typically were able to utilize many hundreds of

Fourier components because of the many different baselines that can be formed with an array of radio telescope dishes. They also benefited from a strong signal-to-noise ratio, a consequence both of the very large collecting area of their Earth-based telescopes and of the very small energy per radio photon. By contrast, X-ray images produced by *RHESSI* had to be constructed from a much more limited number (typically $\sim$100) of sparsely distributed Fourier components. Furthermore, the limited collection area available on a space mission, combined with the relatively high energy per X-ray photon, meant that, despite the relative proximity of the Sun, there were many fewer photons available with which to construct each image. Consequently, the associated Poisson statistics for such low count rates meant that observers had to deal with much noisier data. Combined, these essential differences meant that the extensive analysis software that already existed for radio astronomy provided to be only a starting point for developing new techniques optimized for this new context. To make things more interesting, construction of images from sparse, noisy Fourier data is hardly intuitive, and extensive testing and validation of the methods was necessary to ensure that they produced images with sufficient accuracy and fidelity for solar scientists to employ them in addressing compelling science problems. And to make things more timely, this Fourier-based way of realizing image structure is also at the basis of two other space missions: *STIX* on-board the recently-launched ESA Solar Orbiter and the HXI instrument scheduled for the Chinese Advanced Space-based Solar Observatory.

This book summarizes the results of this development of image reconstruction techniques specifically designed for this form of data. It covers a set of published works that span over two decades, during which it is fair to say that there was very little in the way of a guiding "script." Over this extended period of time, as the various image reconstruction techniques were introduced, developed, validated, and applied to observations, it became more and more apparent to the authors that it would be a good idea to put together a compendium of these methods and their applications, hence the book you are now reading. The order in which the various image reconstruction methods are presented reflects not so much the chronological order of their development but rather the similarities and differences among them, and the degree to which they may (or may not) be useful in addressing the science problems for which they were created.

The book is intended as a reference text for researchers who seek to better understand the scientific context for, and essence of, the various image construction methods appropriate to indirect-imaging instruments like *RHESSI* (and more recently *STIX* and HXI). We hope that it will help them select the best method(s) for the scientific task they wish to address. Students or newcomers to the field should find this book useful as a single reference of the various image reconstruction methods that have been developed, the relative strengths and limitations of each, and the scientific results that have emerged from their application to observations.

Many of the methods described in this book were developed during a series of focused week-long meetings in Bern, Switzerland, sponsored by the International Space Studies Institute (ISSI). We thank that institution for providing us with a quiet and comfortable place where we could focus on these scientific investigations.

We are pleased to thank a number of individuals who have inspired and/or encouraged us to produce this monograph. Various participants, far too numerous to list exhaustively, have engaged in discussions of *RHESSI* hard X-ray images and their scientific consequences at a number of both formal and informal meetings. This includes a series of over 20 specialized *RHESSI* science and data analysis workshops held at various locations (alternating between the US and Europe but also including Nanjing, China in 2011) over the period from 2001 to the present day.[1] We thank all of these individuals for their stimulating input, with especial appreciation to the following people for their unique contributions: Mike Appleby, Markus Aschwanden, Arnold Benz, Steven Christe, André Csillaghy, Martin Fivian, Lindsay Glesener, Hugh Hudson, Eduard Kontar, Säm Krucker, Tom Metcalf (deceased), Pascal Saint-Hilaire, Ed Schmahl, Gerald Share, Albert Shih, David Smith, Kim Tolbert, Frank van Beek, Astrid Veronig, Nicole Vilmer, and Alexander Warmuth.

We particularly acknowledge the unparalleled work of Gordon Hurford, who was largely responsible for the conception, design, and implementation of the *RHESSI* (and subsequently *STIX*) imaging capabilities. His contributions to the realization of these missions have been, and continue to be, crucial at many different levels.

Finally, all four authors recognize that our professional (and personal) lives have benefited enormously from the impressive competencies, and unforgettable friendships, of four individuals who are sadly no longer with us: Reuven Ramaty, a pioneer in the field of gamma-ray astronomy and a highly engaged contributor to providing the scientific justification for *RHESSI* so necessary for success in NASA's highly competitive proposal selection process (the "R" in *RHESSI* honors these contributions); Bob Lin, the original *RHESSI* Principal Investigator, whose vision (and tenacious pursuit of it in the face of a veritable cornucopia of challenges before its eventual launch in February 2002) led to the eventual availability of the unparalleled *RHESSI* data; John Brown, whose pioneering theoretical studies into the relationship between solar hard X-rays and the electrons that produce them set the stage for the *RHESSI* mission; and Richard Schwartz, whose unique combination of technical knowledge and scientific insight was responsible for the transformation of many of the methodologies described in this text into computational algorithms that could profitably be applied to actual data by mere mortals trying to understand the observations in their scientific context.

[1] https://hesperia.gsfc.nasa.gov/rhessi3/news-and-resources/meetings/index.html.

We thank Paolo Massa and Emma Perracchione for proofreading the manuscript and suggesting various insightful comments and corrections. Any inaccuracies, or errors of omission or commission, that remain are, of course, the responsibility of the authors.

Genoa, Italy Anna Maria Massone

Genoa, Italy Michele Piana

Bowling Green, KY, USA A. Gordon Emslie

Greenbelt, MD, USA Brian R. Dennis
October 2021

Contents

Acronyms

AIA	Atmospheric Imaging Assembly; an instrument on SDO
AIPS	Astronomical Image Processing System
ASO-S	Advanced Space-based Solar Observatory; a proposed Chinese mission
AU	Astronomic Unit; a unit of length, the mean distance from Earth to the Sun
BAT	Burst Alert Telescope; an instrument on Swift
CCD	Charge Coupled Device; a type of detector used in digital imaging
CdTe	Cadmium Telluride; material used for STiX X-ray detectors
CHSKP	Carmichael-Hirayama-Sturrock-Kopp-Pneuman; a solar flare model defined by a sequence of influential papers
CME	Coronal Mass Ejection
CMOS	Complementary Metal-Oxide Semiconductor; a type of field-effect transistor
COMPTEL	COMPton TELescope; an instrument on the NASA Compton Gamma-Ray Observatory operational from 1991 to 2000
COSI	Compton Spectrometer and Imager; a SMEX soft gamma-ray survey telescope (0.2–5 MeV)
EM	Expectation Maximization
ESA	European Space Agency
EUV	Extreme Ultra-Violet; electromagnetic radiation
FFT	Fast Fourier Transform; an algorithm for the fast computation of the Fourier transform
FIERCE	Fundamentals of Impulsive Energy Release in the Corona Explorer; a NASA MidEX proposed in 2020
FOXSI	Focusing Optics X-ray Solar Imager; an instrument using focusing optics in the hard X-ray domain, flown on a series of rocket flights and proposed as a NASA SMEX in 2018
FWHM	Full Width at Half Maximum
GCV	Generalized Cross Validation; a technique used to regularize the function recovered from inversion of an integral equation

GOES	Geostationary Operational Environmental Satellites; a series of synoptic monitoring satellites; the intensity of soft X-ray radiation measured by GOES provides a scheme for classifying the intensities of solar flares
GOES-R	Geostationary Operational Environmental Satellites-R Series; a specific series of four satellites in the GOES framework
GRIPS	Gamma-Ray Imager/Polarimeter for Solar flares; an instrument to perform imaging spectro-polarimetry on the hard X-ray and gamma-ray emission from solar flares, flown on a 10-day balloon flight from Antarctica in January 2016
GUI	Graphical User Interface; a user-friendly tool for use with the *RHESSI* software
HPD	Half-Power Diameter
HXI	Hard X-ray Imager; an instrument on the proposed Chinese ASO mission
HXIS	Hard X-ray Imaging Spectrometer; an instrument on SMM; operational from February until December 1980
HXT	Hard X-ray Telescope on Yohkoh
IDL	Interactive Data Language; a sequential programming language by Harris Geospatial Solutions, Boulder, Colorado
IR	Infrared; electromagnetic radiation
IXPE	Imaging X-ray Polarimetry Explorer; a mission to perform imaging spectropolarimetry on soft X-ray emission from astrophysical sources
KKT	Karush-Kuhn-Tucker; a theorem used in the EM method
KL	Kullbach-Leibler; a function used in the EM method
MEM	Maximum Entropy Method; a method used to optimize the output of an image reconstruction algorithm
MidEX	Mid-level Explorer; a type of NASA science-oriented satellite larger than a SMEX
MiSolFA	Micro Solar Flare Apparatus; a proposed small satellite similar in design to STIX
NASA	National Aeronautics and Space Administration
NOAA	National Oceanic and Atmospheric Administration
NuSTAR	Nuclear Spectroscopic Telescope ARray; a NASA SMEX mission launched on June 13, 2012
OSO	Orbiting Solar Observatory; a series of NASA satellites launched in the 1960s, that made the first (spatially unresolved) measurements of X-rays from the Sun
OSPEX	Object SPectral EXecutive; an object-oriented package in IDL for spectroscopy and imaging spectroscopy analysis

PhoENiX	Physics of Energetic and Non-thermal plasmas in the X (= magnetic reconnection) region; a Japanese mission to probe the physics in the magnetic reconnection ("*X*"-point geometry) region in which the primary energy release in solar flares occurs
PSF	Point Spread Function; the response of an instrument to a point source
RAS	Roll Angle System; a system on *RHESSI* to determine roll orientation
RESIK	REntgenovsky Spekrometr Izognutymi Kristalami; an instrument on the Russian CORONAS-F satellite, launched on July 31, 2001
RHESSI	Ramaty High-Energy Solar Spectroscopic Imager; a NASA Small Explorer mission operational from February, 2002 until April, 2018
RMC	Rotating Modulation Collimator; a bi-grid system that modulates the transmission of X-rays as the grid system is rotated; used on *RHESSI*
SAPPHIRE	SolAr Polarimeter for Hard x-Rays; a CubeSat module to extract polarization information from solar X-rays
SAS	Solar Aspect System; a system on *RHESSI* that, when combined with the RAS data, provides precise pointing information
SDAC	Solar Data Analysis Center; serves data from recent and current space-based solar-physics missions, funds and hosts much of the SSW library, and leads the Virtual Solar Observatory (VSO) effort.
SDO	Solar Dynamics Observatory; a NASA mission launched on February 11, 2010
SEE	Solar Eruptive Event; an event that consists of both a solar flare and a coronal mass ejection (CME)
SEPs	Solar Energetic Particles
SHARPIE	The Solar HARd x-ray Polarimer/Imager Experiment; a proposed instrument for solar X-ray imaging spectropolarimetry using SAPPHIRE modules
SMC	Sequential Monte Carlo; a sampling method
SMEX	SMall EXplorer; a type of NASA science-oriented satellite
SMM	Solar Maximum Mission; a NASA satellite operational from February, 1980, until December, 1989
SSW	SolarSoftWare; a collection of IDL codes used extensively in analysis of data from *RHESSI* and other solar instruments
STIX	Spectrometer/Telescope for Imaging X-Rays; an instrument on the ESA Solar Orbiter mission, launched on February 9, 2020
SUVI	Solar Ultraviolet Imager; an instrument on the GOES-R series satellites
Swift	The Neil Gehrels Swift Observatory, a MidEX to study gamma-ray bursts, launched on November 20, 2004
SXT	Solar X-ray Telescope; an instrument on the Japanese Hinotori spacecraft launched on February 21, 1981
TRACE	Transition Region and Coronal Explorer; a NASA SMEX, operational between April, 1998, and June, 2010

URA	Uniformly Redundant Array; a type of coded mask layout used to construct hard X-ray images
URL	Uniform Resource Locator; an address in the World Wide Web
UV	Ultra-Violet; electromagnetic radiation
Yohkoh	Japanese for "Sunbeam," solar spacecraft, operational from September 1991 until December 2001

Chapter 1
Hard X-Ray Emission in Solar Flares

Abstract In this chapter we provide an overview of the solar flare phenomenon with particular emphasis on aspects that can be best investigated through X-ray imaging. Studied for well over a century and a half, the observational manifestations of a solar flare are relatively well cataloged, and its essential nature—the explosive release of energy stored in current-carrying magnetic fields in the outer solar atmosphere—is generally accepted. However, the processes through which this stored energy is converted into the three energetically most important products—hot plasma, erupting material, and accelerated particles (both electrons and ions)—are still poorly understood. High-energy ("hard") X-ray emission provides important information on the accelerated electrons, and the acquisition of imaging spectroscopy information on these hard X-rays is a key element in determining the spatial, temporal, and spectral properties of the electrons, and so to ultimately solving the puzzle of solar flare initiation and development.

1.1 A Brief Overview of Solar Flares

The term "solar flare" was originally used for localized white-light brightenings on the solar disk first seen independently[1] by Carrington [33] and Hodgson [89] during the now-famous "Carrington event" in 1859. Hodgson writes: "While observing a group of solar spots on the 1st September, I was suddenly surprised at the appearance of a very brilliant star of light, much brighter than the sun's surface, most dazzling to the protected eye, illuminating the upper edges of the adjacent spots and streaks, not unlike in effect the edging of the clouds at sunset; the rays extended in

[1] In Carrington's article, he writes, in a model of scientific objectivity: "It has been very gratifying to me to learn that our friend Mr. Hodgson chanced to be observing the sun at his house at Highgate on the same day, and to hear that he was a witness of what he also considered a very remarkable phenomenon. I have carefully avoided exchanging any information with that gentleman, that any value which the accounts may possess may be increased by their entire independence."

M. Piana et al., *Hard X-Ray Imaging of Solar Flares*,
https://doi.org/10.1007/978-3-030-87277-9_1

all directions; and the centre might be compared to the dazzling brilliancy of the bright star α Lyrae when seen in a large telescope with low power."

Flare-associated brightenings are far from confined to the white-light observations associated with the seminal discoveries of 1859; they extend across the whole electromagnetic spectrum, from radio waves with wavelengths in excess of hundreds of kilometers ($> 10^7$ cm) through infra-red (IR), white light, ultraviolet (UV), and X-rays, to gamma-ray emission with wavelengths as short as 10^{-12} cm, corresponding to photon energies in excess of 100 MeV. The detection and analysis of the emission across so many different wavelength bands has been the principal way in which we have come to our current understanding of what solar flares are, and how conditions in the active Sun lead to their production. On one hand, this enormous twenty-orders-of-magnitude wavelength/energy range provides a veritable cornucopia of information on the different physical processes that produce the emission in each range. On the other hand, it presents difficult challenges, since each few decades of wavelength/energy require different detector technologies and an understanding of different radiation mechanisms.

It is now generally accepted that a solar flare and its often-associated coronal mass ejection (CME) (together, a "solar eruptive event" or SEE) are driven by energy stored in the magnetic field of the solar corona. Our current understanding of what a solar flare is, and how it is connected to the CME, is illustrated in Fig. 1.1. Within a so-called "active region" covering a small fraction (typically measured in millionths) of the solar disk, this energy gradually builds up over hours or days until it is suddenly and impulsively released through a process known as *magnetic reconnection*, which occurs when oppositely directed magnetic field lines come together and "reconnect" (i.e., change their connectivity and/or topology [158]). Magnetic reconnection is, by now, a well-known but not fully understood physical process that occurs in highly conducting plasmas in various domains, including terrestrial Tokamak fusion machines, the magnetospheres of the Earth and other planets, in active galaxies, and, most relevant to this book, in the outer atmospheres of the Sun and stars. In a solar flare, the reconnection enables the magnetic field to relax to a lower energy configuration, in the process transferring (in large solar flares) more than 10^{32} ergs of energy (the equivalent of over a billion one-megaton hydrogen bombs) into various other forms, including heating, particle acceleration, and mass motion [66].

The series of events, and the physical processes involved, that follow the primary release of energy through magnetic reconnection are far from fully understood. However, the wealth of available observations does allow us to conclude [185] that the following processes occur in most large solar eruptive events.

1. The hot ionized gas ("plasma") on the reconnecting magnetic field lines is directly heated to temperatures that can exceed 50 million degrees Kelvin (50 MK).
2. Charged particles (electrons, protons, and heavier ions) are accelerated, predominantly downwards on closed magnetic structures ("loops") towards the solar surface, but also upwards on magnetic field lines that may open into

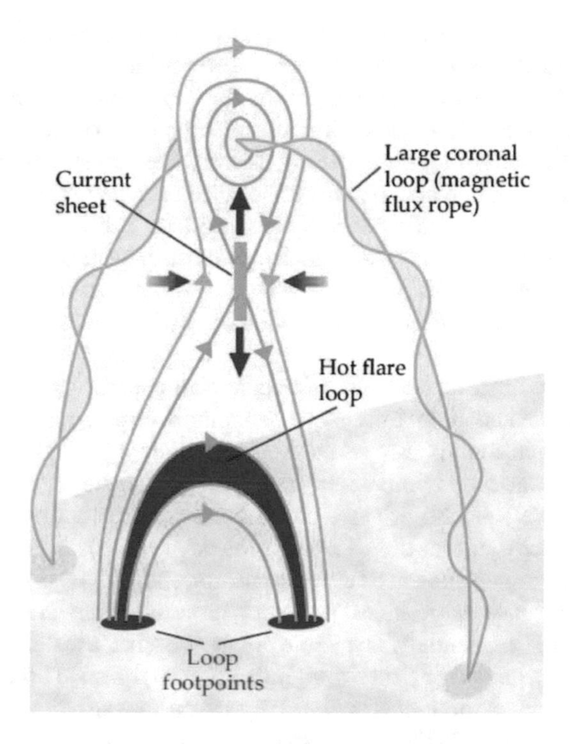

Fig. 1.1 Diagram showing the Standard Model of a solar eruptive event (SEE). Magnetic field lines (shown in blue) emerge from the solar photosphere (yellow) and pass into the corona. Some of them are drawn into a current sheet where magnetic reconnection takes place, allowing energy stored in the magnetic field to be released. Newly reconnected field lines (shown in green) move upwards and downwards away from the reconnection site. Those field lines moving downwards produce the hot flare loops shown in red, while those moving up add to the flux rope (light blue) that connects back to the solar surface. The energy released during this process is transferred to heating the coronal plasma, accelerating charged particles (electrons, protons, and heavier ions), and driving the flux rope away from the Sun to become a coronal mass ejection. The particles accelerated in the downward direction are constrained by electromagnetic forces to spiral along the magnetic field lines until they reach the indicated loop footpoints in the higher density region of the lower corona and upper chromosphere. Here, the accelerated electrons produce the bright hard X-ray sources through the bremsstrahlung process and the ions produce gamma-ray emission through nuclear interactions. [Reproduced from [91], with permission of the American Institute of Physics.]

interplanetary space. The particles are accelerated to energies that greatly exceed the average thermal energy of the particles in a heated plasma ($\sim$2 keV for an electron in a 20 MK plasma) and hence are referred to as being "nonthermal."

3. The accelerated charged particles spiral along the magnetic field lines on which they are produced. Those particles moving upwards may, if they are on open field lines, escape from the Sun and, together with particles accelerated in the shock wave of the associated CME, add to the flux of so-called "solar energetic particles" (SEPs). Those moving downwards on closed field lines will continue

until they reach the so-called "footpoints," where the magnetic loop penetrates the higher density regions of the lower corona and upper chromosphere. There, the ions lose their energy by nuclear interactions to produce gamma-ray emission, while the electrons lose most of their energy by Coulomb collisions with electrons in the ambient plasma. Along the way, the accelerated electrons scatter off ambient protons and heavier ions to produce so-called "hard" X-ray emission through a process known as *bremsstrahlung* (see Sect. 1.2.2). Although only about 10 parts per million of their energy is lost to such bremsstrahlung X-ray emission, this radiation is a key diagnostic of the properties of the accelerated electrons—their location, energy, and pitch-angle distribution about the magnetic field.

4. The passage of this intense beam of accelerated electrons generates an electric field that is created both by charge separation effects and by the changing magnetic field in the vicinity of the beam (see, e.g., [2, 121, 195, 210]). This electric field can be an important factor in the dynamics of the accelerated electrons [42, 59]. It also drives a co-spatial return current of ambient thermal electrons, thereby transporting electrons back to the coronal acceleration site. This continual replenishment of the acceleration site with fresh electrons resolves the so-called "number problem," wherein more electrons are accelerated in a large flare than are initially available in the coronal acceleration site. It has recently been claimed [3] that a significant component of the return current may "run away" as high-energy, hard X-ray producing, electrons traveling in the opposite direction to the beam itself.
5. Through Coulomb collisions with ambient electrons (see Sect. 1.2.3), the accelerated particles heat the chromospheric plasma at the loop footpoints, from initial temperatures of the order of a few *thousand* degrees Kelvin to temperatures in excess of ten *million* degrees Kelvin (10 MK). Because of the relatively high inertia of the dense chromospheric gas, its density stays relatively constant during the initial rapid heating phase, and this combination of near-constant density and rapidly increasing temperature increases the pressure in the lower atmosphere by several orders of magnitude. The heated plasma is forced both upwards and downwards away from this region of enhanced pressure, and, because it is the "path of least resistance," the most significant motion is upwards into the corona—a process that is usually (but erroneously) referred to as "chromospheric evaporation" ([4], and references therein). The material travels upwards [128] at speeds of up to the local sound speed of several hundred km s^{-1}, taking a few tens of seconds to reach the loop top, where it mingles with the plasma directly heated during the magnetic reconnection process to create the hot, dense, loops shown in Fig. 1.1. These loops eventually cool by thermal conduction of energy back down to the cooler chromosphere and by soft X-ray and Extreme Ultra-Violet (EUV) radiation to outer space, becoming the post-flare loops that can still be seen [161] in UV radiation for hours after the main phase of the flare.
6. Field lines that are driven upward from the reconnection site add to the overlying magnetic "flux rope" that is driven away from the Sun to become the CME.

The bulk of the radiative emission from a flare [66] is from material heated to temperatures that range from chromospheric values of a few thousand K to coronal plasma with temperatures that can exceed 50 MK in the biggest flares. Because of the huge amount of optical radiation produced from the so-called "quiet" Sun, excess flare emission is not commonly visible in white light (the exception, of course, being in unusually large events such as the one [33, 89] that formed the seminal discovery of the flare phenomenon). Instead, flare emission is most conspicuous in wavelength ranges for which it represents a significant excess over that from the quiet Sun, i.e., in the UV, EUV, and soft X-ray ranges (i.e., at energies of up to ~10 keV or wavelengths down to ~1 Å).

One of the overarching goals of solar flare research is to understand the nature and causes of impulsive energy release in the active Sun, and it must be remarked that this is not only for its intrinsic scientific value, but is also motivated by the need to develop a reliable capability to predict, and hence mitigate, the occurrence and severity of solar eruptive events, with their associated hazardous Space Weather [14] effects at the Earth. Consequently, much effort has been expended, both observationally and theoretically, toward understanding the emission produced from the material energized by the flare, both in order to understand the mechanisms through which this energized material is produced, and to use it as a means to diagnose the environment in which the flare energy release occurs.

The majority of the heated (and indeed, much of the pre-heated) material in flares is in the *plasma* state, with the atoms partially or completely stripped of their electrons. This plasma rapidly reaches what is called thermal equilibrium, in which the ions and the electrons have Maxwellian phase-space distributions (particles cm^{-3} (cm s^{-1})$^{-3}$) for the density of particles per unit volume with velocities within an infinitesimal element d^3v of three-dimensional velocity space:

$$f(v)\, \mathrm{d}^3 v = \left(\frac{m}{2\pi k_B T} \right)^{3/2} n\, e^{-mv^2/2k_B T}\, \mathrm{d}^3 v\,. \tag{1.1}$$

Here n (cm^{-3}) is the local number density, m is the particle mass, k_B is Boltzmann's constant and T is the thermodynamic temperature. The corresponding energy distribution (particles cm^{-3} per unit energy E) is given, for isotropic distributions, by the relation $f_E(E)\, dE = f(v)\, 4\pi v^2\, dv$, with $E = \frac{1}{2} m v^2$, resulting in

$$f_E(E)\, dE = \sqrt{\frac{2}{\pi (k_B T)^3}}\, n\, E^{1/2}\, e^{-E/kT}\, dE\,. \tag{1.2}$$

The exponential term in Eq. (1.2) means that there are very few particles with energies $\gtrsim 3k_B T$, which for 10 MK plasma is $\simeq$3 keV (1 keV $\equiv$ 11.6 MK). However, during the early impulsive phase of most flares, the evolving magnetic and electric fields associated with the magnetic reconnection produce large fluxes of both electrons and ions with energies much higher than $3k_B T$. The spectra of these particles generally fall off with photon energy according to a power-law functional form that is flatter (more gradual) than the above exponential thermal expression; they are thus referred to as "nonthermal" particles. As discussed in [66], nonthermal particles can contain an amount of energy that is comparable to that in the thermal plasma. Accelerated electrons in particular are manifested by the variety of radiation signatures that they produce, including radio emission from gyrosynchrotron and plasma kinetic processes (e.g., [5, 137, 138]), and hard X-ray *bremsstrahlung*.

1.2 Hard X-Ray Emission from Flares

1.2.1 Acceleration of Nonthermal Electrons

Magnetic reconnection intrinsically involves changes in the magnetic field **B**, and hence in the creation of electric fields, $\mathcal{E}$, resulting both from local changes in the magnetic field **B** given by Faraday's Law:[2]

$$\nabla \times \mathcal{E} = -\frac{1}{c}\frac{\partial \mathbf{B}}{\partial t} \tag{1.3}$$

and from advection (via the transverse flow of a fluid at velocity v):

$$\mathcal{E} = -\frac{v}{c} \times \mathbf{B}\,, \tag{1.4}$$

where $c = 3 \times 10^{10}$ cm s^{-1} is the speed of light. These electric fields act on any charged particles present and can, under appropriate conditions, accelerate them to very high energies. There is a rich array of processes by which electric fields are produced in solar flares [212], and a review of these processes is outside the scope of this volume. Nevertheless, it is worth exploring briefly the threshold value of the electric field that is required to produce substantial particle acceleration, particularly of electrons.

Consider an electron of mass m_e (grams) and charge magnitude e (measured in statcoulombs) subject to both an applied electric field $\mathcal{E}$ (statvolt cm^{-1}) and a friction (drag) force due to collisions with ambient particles. Because high-velocity particles spend less time in the Coulomb field of an ambient particle, the drag force experienced is, unlike for "ordinary" drag forces such as air resistance, a *decreasing* function of velocity. The drag force can be measured in terms of the *collision frequency* ν_c, the temporal rate (s^{-1}) associated with the change of a physical quantity such as energy E or speed v: e.g., $|dE/dt| = \nu_c\,|E|$. It can be

[2] In this text, we follow astronomical convention and use electrostatic cgs units, where 1 statvolt $\simeq$300 V. Thus, the electric field has the units of statvolt cm^{-1} and **B** has the same units as $\mathcal{E}$. The above form of Faraday's law reflects this choice of units. However, particle energies will be frequently stated in keV units.

shown (e.g., [36]) that $\nu_c(v) = \nu_o\,(v_e/v)^3$, where ν_o is the collision frequency for an electron moving at the thermal speed $v_e = \sqrt{2k_B T/m_e}$. The equation of motion (in one dimension, for simplicity) is thus

$$m_e \frac{dv}{dt} = e\,\mathcal{E} - \nu_o \left(\frac{v_e}{v}\right)^3 m_e\, v\,, \tag{1.5}$$

or

$$\frac{dv}{dt} = \frac{e\,\mathcal{E}}{m_e}\left[1 - \left(\frac{\mathcal{E}_{\mathcal{D}}}{\mathcal{E}}\right)\left(\frac{v_e}{v}\right)^2\right], \tag{1.6}$$

where the *Dreicer* [54] field

$$\mathcal{E}_D = \frac{\nu_o\, m_e\, v_e}{e} = \frac{8\pi\, n\, e^3 \Lambda}{m_e\, v_e^2} = \frac{4\pi\, e^3 \Lambda}{k_B}\,\frac{n}{T} \tag{1.7}$$

and Λ is the *Coulomb logarithm* (e.g. [36], [180]). Substituting numerical values $e = 4.8 \times 10^{-10}$ statcoulomb and $k_B = 1.38 \times 10^{-16}$ erg K^{-1}, and taking $\Lambda \simeq 20$, gives $\mathcal{E}_D \simeq 2 \times 10^{-10}\,(n/T)$ statvolt $\mathrm{cm}^{-1} = 6 \times 10^{-8}\,(n/T)$ V cm^{-1}. Typical values for the flaring solar corona are $n \simeq 10^{10}$ cm^{-3} and $T \simeq 10^7$ K, so that $\mathcal{E}_D \simeq 10^{-4}$ V cm^{-1}, corresponding to a voltage difference of about 100 keV over the 10^9 cm length of an active region loop. As will be seen shortly below, electrons of such an energy emit photons in the so-called "hard X-ray" domain.

If an electron has a velocity $v < v_e\,(\mathcal{E}_{\mathcal{D}}/\mathcal{E})^{1/2}$, then the right side of Eq. (1.6) is negative, and the particle slows down. Since the drag force increases with decreasing velocity (according to Eq. (1.5), it is proportional to $1/v^2$), the electron decelerates at an increasing rate and is rapidly absorbed into the background Maxwellian distribution. On the other hand, if an electron has a velocity $v > v_e\,(\mathcal{E}_{\mathcal{D}}/\mathcal{E})^{1/2}$, the right side of Eq. (1.6) is positive and the electron accelerates. The resulting increase in velocity leads to an even smaller collisional drag force, and the net force on the electron increases further. This leads to the phenomenon known as *runaway acceleration*. Since the threshold velocity for runaway acceleration scales as $(\mathcal{E}/\mathcal{E}_{\mathcal{D}})^{-1/2}$, the fraction of electrons that suffer runaway acceleration depends on the strength of the electric field with respect to the Dreicer field. For sub-Dreicer fields $\mathcal{E} \ll \mathcal{E}_D$, runaway acceleration occurs only for the small fraction of electrons in the high-velocity "tail" of the Maxwellian distribution (1.1). However, for electric fields $\mathcal{E} > \mathcal{E}_D$, a significant fraction of the distribution undergoes runaway acceleration. Coulomb collisions then lead to a fairly prompt repopulation of the depleted high-energy tail of the distribution, so that essentially *all* of the electrons end up being accelerated to hard-X-ray-emitting energies of tens to hundreds of keV.

1.2.2 Hard X-Ray Production by Accelerated Electrons: The Bremsstrahlung Process

When an accelerated electron of energy E encounters an ambient proton (or heavier ion) in the ionized plasma environment of the solar atmosphere, the Coulomb force between them deflects the electron.[3] This acceleration of the electron throughout the duration of its encounter with the proton causes it to radiate away some of its energy; a fully relativistic quantum mechanical analysis (e.g., [87]) shows that a single photon is produced as a result of this encounter. This photon emission process is known as *bremsstrahlung*, a German word formed from the words for "braking" and "radiation." Bremsstrahlung photons have a range of energies up to that of the exciting electron, which for the electrons accelerated in solar flares is, as we noted above, in the so-called "hard" X-ray energy range from $\sim$10 to $\sim$100 keV.

Hard X-ray emission cannot penetrate the Earth's atmosphere to be observed by ground-based instruments; however, solar flare bremsstrahlung has been observed by space-borne instruments since the 1960s. Qualitatively, hard X-ray emission provides unambiguous evidence for the high-energy electrons that produce them; quantitatively, the properties of the observed bremsstrahlung radiation—its time history, energy spectrum, and location of origin—can be used to infer properties of the electrons—their time history, spectrum, and location.

It is especially noteworthy that the emitted bremsstrahlung photons, being uncharged, are unaffected by the magnetic fields that bend the trajectories of charged particles traveling from the Sun. Further, the amount of intervening material along their trajectory to the Earth is too low to cause any significant scattering or absorption. Consequently, each X-ray bremsstrahlung photon travels essentially unhindered in a straight line at the speed of light, directly from the electron/proton interaction that produced it to the detection point within an instrument in Earth orbit, and with all the properties that it possessed at production remaining intact. The properties of the X-ray bremsstrahlung photons (spectrum, direction [i.e., location], directivity, and polarization) thus provide precise information about the accelerated electrons at the Sun.

The deduction of electron properties from spacecraft observations of the hard X-ray bremsstrahlung that they produce is the overarching thrust behind the research described in this book. It focuses on observations of the spectral and spatial distributions of the hard X-ray emission in solar flares, and hence the implied properties of the accelerated electrons—how and where they are accelerated, their energy distribution (spectrum), and how they propagate in, and interact with, the ambient solar material to produce many of the manifestations of a solar flare. The knowledge gained from observations of this solar "laboratory," situated at mere 8

[3] The amount of momentum transferred to the proton by the electron is, of course, equal and opposite to the amount of momentum transferred to the electron by the proton. However, because the proton mass is 1836 $\times$ the electron mass, the proton is relatively undisturbed by this momentum transfer.

light *minutes* from Earth (as compared to many light *years* for other astrophysical sources), can be applied to increase our understanding of high-energy processes in the rest of the Universe, including in sources such as blazars [73] and active galactic nuclei [171], which are situated at such distances (many light years) that direct observation of the spatial structure of the high-energy radiation that they produce is impossible.

In general, the observed photon intensity $J(\epsilon; \Omega)$ (photons cm^{-2} s^{-1} keV^{-1} sr^{-1}, differential in photon energy and solid angle Ω) depends in a complicated way on the energy E and orientation Ω' of the exciting electrons and on the polarization state of the emitted photon (see the Appendix in [132], and references therein). However, a simpler, but nevertheless useful, exercise is to consider the total hard X-ray emissivity (photons cm^{-2} s^{-1} keV^{-1}) at photon energy ϵ, averaged over all emission directions Ω. Roughly, this is proportional to the magnitude of the electron flux $\int_\epsilon^\infty F(E)\,dE = \int_\epsilon^\infty v(E)\,f_E(E)\,dE$ (cm^{-2} s^{-1} above energy ϵ) and to the number of ambient ions that are encountered by the electrons, expressed as a volume integral $\mathcal{N} = \int n(\mathbf{r})\,dV$ of the ambient number density n (cm^{-3}). The observed emission is also inversely proportional to R^2, where R is the distance between the source and the observer, so that $I(\epsilon) \propto (\mathcal{N}/R^2) \int_\epsilon^\infty F(E)\,dE$. Dimensional analysis considerations show that $I(\epsilon)$ must also be proportional to a quantity $Q(\epsilon, E)$, with dimensions cm^2 keV^{-1}. This quantity is termed the *bremsstrahlung cross-section*; it basically quantifies the likelihood (per unit photon energy) that an electron of energy E will cause the emission of a photon of energy ϵ.

A formula for the emitted intensity $J(\epsilon)$ that takes all pertinent variables into account has been provided in equation (1.2) of [111]; see also [132]. However, it is instructive for the present purposes to consider the simpler case of an isotropic cross section $Q(\epsilon, E)$, for which the radiation flux (photons cm^{-2} s^{-1} per unit photon energy ϵ, observed by a detector located at a distance R from the source), is

$$I(\epsilon) = \frac{1}{4\pi R^2} \int_\epsilon^\infty \int_V n(\mathbf{r})\, F(E, \mathbf{r})\, Q(\epsilon, E)\, dE\, dV\,. \tag{1.8}$$

Rigorous, fully relativistic, formulae for the bremsstrahlung cross-section $Q(\epsilon, E)$ are provided in [105]; an approximate non-relativistic form, that is useful in various analytic derivations (see, e.g., Eq. (1.23) below), is the Kramers cross-section

$$Q_K(\epsilon, E) = \frac{8\alpha}{3}\, r_e^2\, \frac{m_e c^2}{\epsilon\, E}\,, \tag{1.9}$$

which well represents the general (inverse) dependencies on both the photon energy ϵ and the electron energy E (so that, for a given electron energy E, the probability of bremsstrahlung emission increases at lower emitted photon energies ϵ; and for a given photon energy ϵ, the probability of its production by an electron of energy E

decreases with E). A significant improvement [23] over the Kramers cross-section is the Bethe-Heitler cross-section

$$Q_{BH}(\epsilon, E) = \frac{8\alpha}{3} r_e^2 \frac{m_e c^2}{\epsilon E} \ln \frac{1 + \sqrt{1 - \epsilon/E}}{1 - \sqrt{1 - \epsilon/E}} , \tag{1.10}$$

which, unlike the Kramers cross-section, has the desirable property of vanishing at the bremsstrahlung threshold $E = \epsilon$. (In the above equations, $\alpha = e^2/\hbar c \simeq 1/137$ is the fine structure constant, $\hbar \simeq 1.05 \times 10^{-27}$erg s is Planck's constant, and $r_e = e^2/m_e c^2$ (cm) is the electron radius.)

Equation (1.8) can be very conveniently written as [29]

$$I(\epsilon) = \frac{1}{4\pi R^2} \int_{\epsilon}^{\infty} \langle nVF(E) \rangle \, Q(\epsilon, E) \, dE \ , \tag{1.11}$$

where

$$\langle nVF(E) \rangle = \int_V n(\mathbf{r}) \, F(E, \mathbf{r}) \, dV \tag{1.12}$$

is the *mean source electron flux spectrum* (in units of cm^{-2} s^{-1} keV^{-1}), the value of the electron flux spectrum integrated over the emitting region, weighted by the density of ambient ions. Equation (1.11) shows that the observed hard X-ray spectrum $I(\epsilon)$ is proportional to the integral of the mean source electron flux $\langle nVF(E) \rangle$, weighted by the cross-section $Q(\epsilon, E)$. It should be noted that since the photon energy ϵ appears as the lower limit of the integral in Eq. (1.11), and since both $\langle nVF(E) \rangle$ and $Q(\epsilon, E)$ are non-negative quantities, then $I(\epsilon)$ must be a monotonically decreasing function of ϵ.

Mathematically, Eq. (1.11) takes the form of a Fredholm integral equation with a *source function* $\langle nVF(E) \rangle$, a *data function* $I(\epsilon)$ and a *kernel* $Q(\epsilon, E)$. This equation can, at least in principle, be formally inverted to yield knowledge of the source function $\langle nVF(E) \rangle$, given the observed data function $I(\epsilon)$ and knowledge of the form of the cross-section[4] $Q(\epsilon, E)$. Such inverse problems are, however, notoriously ill-posed (see, e.g., [44]), so that the unavoidable presence of noise in the data $I(\epsilon)$ renders it very difficult (or even—see Sect. 4.1—impossible) to determine the exact form of the source function $\langle nVF(E) \rangle$. Methods which impose reasonable ("regularized") constraints on $\langle nVF(E) \rangle$ in order to allow its extraction from Eq. (1.11) have been devised and thoroughly tested (e.g., [30, 31, 107–109, 150, 156]).

Although the amount of energy emitted as hard X-rays is a negligible [23] fraction ($\simeq 10^{-5}$) of the energy released in a solar flare, the observed hard X-ray spectrum provides direct diagnostic information on the mean source electron

[4] It is important to note that the result of this inversion to obtain $\langle nVF(E) \rangle$ depends *only* on the form of the cross-section $Q(\epsilon, E)$; no other physics is involved.

spectrum $\langle nVF(E)\rangle$ in the region under observation, requiring only knowledge of the cross-section kernel in Eq. (1.11). For early instruments that measured only the hard X-ray spectrum from the entire field of view, the region V that appears in the mean source electron spectrum $\langle nVF(E)\rangle$ was of necessity the entire flare volume. Later instruments have been able to image the hard X-ray emission on the plane of the sky, so that the pertinent volume becomes $V = \int A(\ell)\, d\ell$, the integral along the line of sight through the flaring volume of the area (on the plane of the sky) of the source at line-of-sight location ℓ.

1.2.3 Relation of the Mean Source Electron Spectrum to the Accelerated Spectrum

The mean source electron spectrum $\langle nVF(E)\rangle$ is exactly that—the average value of the electron flux spectrum $F(E)$ (weighted by the ambient ion density) throughout the observed region. If there is negligible variation in the electron flux spectrum $F(E)$ throughout the bremsstrahlung-emitting region, then $\langle nVF(E)\rangle$ is simply equal to $F(E) \times \int n\, dV$; in this *thin target* situation, the (uniform) electron flux spectrum has the same shape as the mean source electron spectrum that is determined by inversion of Eq. (1.11). However, it is generally accepted that high-energy electrons in solar flares are not necessarily produced throughout the entire flare volume; rather they are accelerated by electric fields in the region of magnetic reconnection and then propagate into other parts of the flaring region. Thus, there will be significant variation in $F(E)$ throughout the bremsstrahlung source and so the mean source electron spectrum will *not* have the same shape as the *injected* (or accelerated) electron spectrum, the quantity that is most directly related to the release of energy in the flare. Such a *thick target* situation requires that we determine the relation between the mean source electron spectrum $\langle nVF(E)\rangle$ and $\dot{N}_o(E_o)$ (s^{-1} keV^{-1}), the rate (per unit injected electron energy E_o) at which electrons are injected into the bremsstrahlung-emitting target.

To determine this important relationship, we follow the methodology of [26]. If an electron of energy E loses energy at a rate dE/dt (keV s^{-1}), then the energy loss rate per unit *column density* N (cm^{-2}; particles per unit area integrated along the trajectory s) is

$$\frac{dE}{dN} = \frac{1}{n(s)\, v(E)} \frac{dE}{dt} \; . \tag{1.13}$$

The number of bremsstrahlung photons per unit photon energy ϵ emitted by an electron of injected energy $E_o$along its path through the target is

$$\mathcal{N}(\epsilon, E_o) = \int_0^{\tau(E_o)} n(s[t])\, v(E[t])\, Q(\epsilon, E[t])\, dt \; , \tag{1.14}$$

where $\tau(E_o)$ is the time at which the energy of an electron with initial energy E_o is reduced to the bremsstrahlung threshold value $E = \epsilon$, so that no further emission at photon energy ϵ can occur. Using Eq. (1.13), noting that $E = E_o$ at $t = 0$ and $E = \epsilon$ at $t = \tau$, this may be written as

$$\mathcal{N}(\epsilon, E_o) = \int_0^\tau \frac{Q(\epsilon, E)}{dE/dN} \frac{dE}{dt} \, dt = \int_\epsilon^{E_o} \frac{Q(\epsilon, E)}{|dE/dN|} \, dE \ . \tag{1.15}$$

Multiplying $\mathcal{N}(\epsilon, E_o)$ (photons keV^{-1}) by the injection rate per unit energy $\dot{N}_o(E_o)$ (electrons s^{-1} keV^{-1}), integrating over all values of E_o, and dividing by the illuminated area gives the observed bremsstrahlung intensity (photons cm^{-2} s^{-1} keV^{-1}) at distance R:

$$\begin{aligned} I(\epsilon) &= \frac{1}{4\pi R^2} \int_\epsilon^\infty \dot{N}_o(E_o) \, \mathcal{N}(\epsilon, E_o) \, dE_o \\ &= \frac{1}{4\pi R^2} \int_\epsilon^\infty \dot{N}_o(E_o) \, dE_o \int_\epsilon^{E_o} \frac{Q(\epsilon, E)}{|dE/dN|} \, dE \\ &= \frac{1}{4\pi R^2} \int_\epsilon^\infty \frac{Q(\epsilon, E)}{|dE/dN|} \, dE \int_E^\infty \dot{N}_o(E_o) \, dE_o \ , \end{aligned} \tag{1.16}$$

where in the last equality we have (carefully!) reversed the order of integration. The observed hard X-ray spectrum $I(\epsilon)$ is thus, quite reasonably, proportional to the rate at which electrons are injected and to the magnitude of the bremsstrahlung cross-section, but inversely proportional to the rate at which electrons lose energy in the target.

There are numerous mechanisms through which electrons can lose (or, in some situations, gain) energy in a magnetized plasma. These include large-scale electric fields [211], interaction with a wide variety of plasma waves [190], synchrotron radiation [148], inverse Compton radiation [37], and, of course, the bremsstrahlung radiation process itself. However, for the electrons that produce hard X-ray emission in solar flares the dominant energy loss mechanism is generally taken to be due to Coulomb collisions (binary elastic collisions between two charged particles interacting through their mutual electric fields) with ambient electrons and ions. Losses due to collisions with ambient protons or heavier ions are much less important than losses to ambient electrons [58]); thus the dominant energy loss mechanism is electron-electron collisions, for which

$$\frac{dE}{dN} = -\frac{K}{E} \ , \tag{1.17}$$

where the constant $K = 2\pi e^4 \Lambda$ ($= 2.6 \times 10^{-18}$ ($\Lambda/20$) cm^2 keV^2), with e (esu) the electronic charge and Λ the Coulomb logarithm [180]. Substituting Eq. (1.17) into Eq. (1.16) gives

$$I(\epsilon) = \frac{1}{4\pi R^2} \frac{1}{K} \int_{\epsilon}^{\infty} E\, Q(\epsilon, E)\, dE \int_{E}^{\infty} \dot{N}_o(E_o)\, dE_o\,. \tag{1.18}$$

Comparison of Eq. (1.18) with the basic equation (1.11) that defines the mean source electron spectrum gives the relation between the injected spectrun $\dot{N}(E_o)$ and the means source electron spectrum $\langle nVF(E) \rangle$ in a collisional thick target:

$$\langle nVF(E) \rangle = \frac{E}{K} \int_{E}^{\infty} \dot{N}_o(E_o)\, dE_o\,. \tag{1.19}$$

We stress that this relation depends only on the physics of the electron interaction, and not on the form of the cross-section $Q(\epsilon, E)$ used to determine the mean source electron spectrum $\langle nVF(E) \rangle$ from observations of the hard X-ray spectrum $I(\epsilon)$. Differentiating Eq. (1.19) with respect to E explicitly gives the injected rate spectrum in terms of $\langle nVF(E) \rangle$:

$$\dot{N}_o(E_o) = K \left[-\frac{d}{dE} \frac{\langle nVF(E) \rangle}{E} \right]_{E=E_o}\,. \tag{1.20}$$

It is important to recognize what is—and what is not—included in the steps involved in deducing the injected rate spectrum $\dot{N}_o(E_o)$ from the observed hard X-ray spectrum $I(\epsilon)$:

1. The first step (Eq. (1.11)) involves inference of the mean source electron spectrum $\langle nVF(E) \rangle$ from the observed hard X-ray spectrum $I(\epsilon)$. It involves *only* knowledge of the form of the bremsstrahlung cross-section $Q(\epsilon, E)$ and does *not* require any knowledge of the physical processes affecting the energy evolution of the electrons in the bremsstrahlung target. Mathematically, it involves the inversion of an ill-posed Fredholm integral equation, with the associated well-known challenges;
2. The second step (Eq. (1.20)) involves inference of the injected rate spectrum $\dot{N}_o(E_o)$ from the mean source electron spectrum $\langle nVF(E) \rangle$. Exactly complementary to the first step, this step does *not* depend on the form of the cross-section $Q(\epsilon, E)$, instead depending *only* on the nature of the physical processes affecting the energy evolution of the electrons in the target. Mathematically, it involves an explicit differentiation (rather than solution of an integral equation), which may also require some form of smoothing to produce physically meaningful results.

As an example of the application *in principle* of the two-step process outlined in Eqs. (1.11) and (1.20) above, we consider an idealized power-law hard X-ray spectrum

$$I(\epsilon) = I_o \left(\frac{\epsilon}{\epsilon_o} \right)^{-\gamma} . \tag{1.21}$$

Step 1: Determine the Mean Source Electron Spectrum

Using Eq. (1.11) with the hard X-ray spectrum (1.21) and the approximate Kramers cross-section (1.9)

$$Q(\epsilon, E) \equiv Q_K(\epsilon, E) = \frac{Q_o}{\epsilon E} \tag{1.22}$$

(where $Q_o = (8\alpha/3)\, r_e^2\, m_e c^2$ has the units cm^2 keV), gives

$$I_o\, \epsilon_o^\gamma\, \epsilon^{1-\gamma} = \frac{Q_o}{4\pi R^2} \int_\epsilon^\infty \frac{\langle nVF(E) \rangle}{E}\, dE . \tag{1.23}$$

Differentiating both sides with respect to ϵ gives the mean source electron spectrum

$$\langle nVF(E) \rangle = (\gamma - 1)\, \frac{4\pi R^2\, \epsilon_o^\gamma}{Q_o}\, I_o\, E^{1-\gamma} , \tag{1.24}$$

which we note is also a power-law, with a spectral index $\gamma - 1$, i.e., one power harder than the photon spectrum.

Step 2: Determine the Electron Injection Rate Spectrum

Substituting the mean source electron spectrum (1.24) into Eq. (1.20) immediately gives the form of the injected rate spectrum:

$$\dot{N}_o(E_o) = \gamma\, (\gamma - 1)\, \frac{4\pi R^2\, K\, \epsilon_o^\gamma}{Q_o}\, I_o\, E_o^{-\gamma-1} . \tag{1.25}$$

This is again a power-law, with a spectral index $\gamma + 1$, i.e., one power *softer* than the photon spectrum, and two powers softer than the mean source electron spectrum (1.24). Physically, this difference of two between the power-law indices for the injected and mean source electron spectra arises from the form of the energy loss rate (1.17), which can be written as $d \ln E/dN = -2K/E^2$, showing that the energy loss cross-section scales as E^{-2}. Consequently, higher energy electrons suffer an energy loss rate that is smaller (by two powers of E) than lower energy electrons, which leads to a spectral flattening characterized by the same two-powers-of-E difference.

The total injected rate $\mathcal{N}$ (s^{-1}) and power P (keV s^{-1}), both above reference energy $E_{\rm ref}$, that correspond to the injected rate spectrum (1.25) are

$$\mathcal{N}(E_{\rm ref}) = \int_{E_{\rm ref}}^{\infty} \dot{N}_o(E_o)\, dE_o = (\gamma - 1)\, \frac{4\pi R^2 K}{Q_o}\, I_o \left(\frac{E_{\rm ref}}{\epsilon_o}\right)^{-\gamma} \tag{1.26}$$

and

$$P(E_{\rm ref}) = \int_{E_{\rm ref}}^{\infty} \dot{N}_o(E_o)\, E_o\, dE_o = \gamma\, \frac{4\pi R^2 K}{Q_o}\, I_o\, E_{\rm ref} \left(\frac{E_{\rm ref}}{\epsilon_o}\right)^{-\gamma} . \tag{1.27}$$

If the energy $E_{\rm ref}$ is taken to be a physical lower limit E_c to the injected electron spectrum,[5] then $\mathcal{N}(E_c)$ and $P(E_c)$ represent the *total* injection rate and power, respectively. Since observed values of the hard X-ray power-law spectral index γ can be quite large (often $\gtrsim 5$), it follows that the total injected rate and power are sensitively dependent on the assumed cutoff E_c (or, more generally, on the form of the injected spectrum at low energies). Of course, rather than simply assuming such an abrupt cutoff at energy E_c (or even a smooth functional form for the low-energy behavior of $\dot{N}_o(E_o)$), we should instead let the observations *reveal* the form of $\dot{N}_o(E_o)$ (and even the veracity of the collisional energy loss expression (1.17)). This will be one of the main goals addressed in Chap. 7, where we apply results obtained from analysis of hard X-ray images of solar flares in order to determine empirically both the form of the accelerated electron spectrum and the manner in which the accelerated electrons lose energy in the solar target.

1.3 History of Solar Hard X-Ray Imaging Observations

Spatially unresolved measurements of hard X-ray emission from the Sun have been made ever since the NASA Orbiting Solar Observatory (OSO) series of satellites, beginning in the 1960s and continuing for well over a decade thereafter [70]. A variety of other instruments, from many nations, have continued this legacy of full-Sun hard X-ray measurements to this day.

The history of hard X-ray *imaging* of solar flares dates back to the early 1980s, but only four instruments have flown that had the capability to image flares with sufficiently fine angular resolution of better than $\sim 10''$. (Here the term

[5] If $E_{\rm ref} = E_c$ is taken to be an actual low-energy cutoff in the injected electron spectrum, then the observed hard X-ray spectrum will not be of the assumed power-law form (1.21). The form of $I(\epsilon)$ in such a case can be found by substituting the form of $\dot{N}_o(E_o)$, with a low-energy cutoff E_c applied, into Eq. (1.19) and then substituting the resulting form of $\langle n V F(E) \rangle$ into Eq. (1.11). Doing so readily verifies that this results in a bremsstrahlung spectrum that is an approximate power law $\epsilon^{-\gamma}$ above $\epsilon = E_c$ and rolls over toward an ϵ^{-1} form at low energies.

"angular resolution" is defined as the angular separation in arcseconds[6] between two sources that the instrument can resolve.) The first two instruments were the Hard X-ray Imaging Spectrometer (HXIS) [193] on NASA's Solar Maximum Mission (SMM), launched on 1980 February 14, and the Solar X-ray Telescope (SXT) on the Japanese Hinotori spacecraft [68], launched on 1981 February 21. These instruments were followed in the 1990s by the Hard X-ray Telescope (HXT, [113]) on the Japanese Yohkoh mission (1991–2001) and in the 2000s by NASA's *Reuven Ramaty High Energy Solar Spectroscopic Imager (RHESSI)* Small Explorer (SMEX) mission [125] (2002–2018).

More recently, the Nuclear Spectroscopic Telescope ARray (NuSTAR, [85]), an astrophysics mission with focusing optics capable of $\sim 10''$ (full width at half maximum - FWHM) imaging in X-rays up to 79 keV, has made intermittent solar observations (e.g., [78]) since its launch on 2012 June 13. Unfortunately, it is effectively limited to energies below $\sim$10 keV for solar flare observations because of the high dead times associated with the instrument's super-high sensitivity down to $\sim$3 keV. The Focusing Optics X-ray Solar Imager (FOXSI, [119]) is also designed to image up to energies as high as 80 keV with order-of-magnitude improvements in sensitivity and dynamic range over *RHESSI*. Furthermore, unlike NuSTAR, it incorporates features designed to prevent saturation by the intense soft X-ray fluxes during larger solar flares. Several versions of the basic FOXSI design have been flown on relatively short-duration balloon and sounding-rocket flights, but to date have only been able to observe small flares at energies below $\sim$10 keV (e.g., [11]).

We will now summarize the main characteristics of the four main instruments with illustrative examples of images made by each. This will show the evolution of capabilities over the four decades and three solar cycles that have elapsed since the first hard X-ray images were made in 1980.

SXT on Hinotori (Japanese for "firebird") used two bi-grid modulation collimators with FWHM resolutions of $30''$ and $38''$ rotating at 4.3 rpm to make images of solar hard X-ray bursts between 17 and 40 keV, with a claimed FWHM response as fine as $7''$. Unfortunately, little information on the observations made with this instrument has been preserved online. It was reported in [184] that, except for a double source seen during a flare on 1981 October 15, most of the 25–50 keV images for nine flares show single compact or extended sources. An example is shown in Fig. 1.2.

HXIS on SMM achieved its $8''$ imaging capability in six energy bands from 3 to 30 keV using simple collimation through a series of holes in ten metal plates stacked one above the other. Images were built up from a total of 432 contiguous square regions of the Sun covering a 6.35-arc-minute field of view with $32''$ pixels

[6] Throughout this book we use both the term "arcsecond" and the conventional astronomical notation $''$ to express both small angles (e.g., in instrument design) and distances within solar sources (for a solar source viewed from the Earth, $1'' \simeq 725$ km on the plane of the sky). The former notation is necessary because (1) we will frequently use the word "arcsecond" in the adjectival sense, and (2) we will make extensive use of quantities defined in inverse angular space (arcsec^{-1}), for which the notation $''^{-1}$ would simply be too awkward.

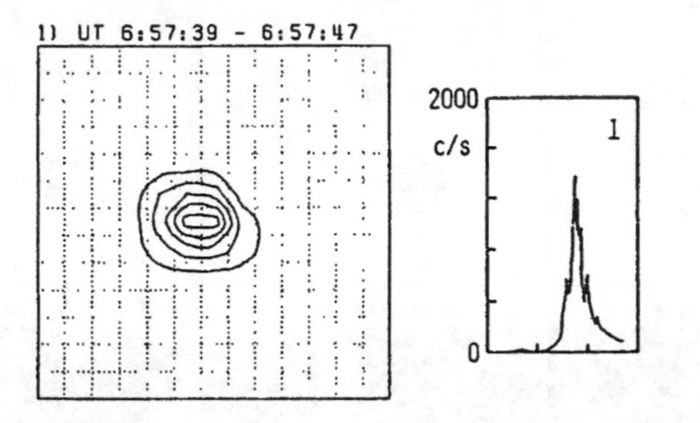

Fig. 1.2 *Left:* Hinotori/SXT 25–50 keV image of a flare on 1981 August 22 showing a single extended source; *Right:* the 30–40 keV count-rate light curve of the flare from 06:56 to 06:59 UT. After [184], used with permission of Springer.

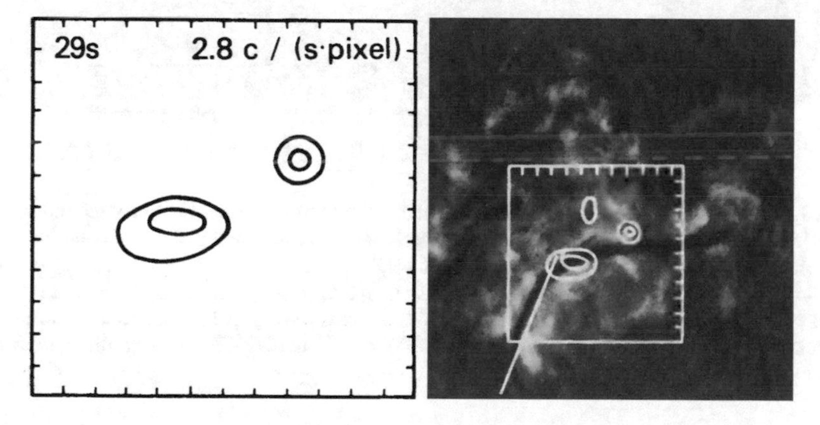

Fig. 1.3 HXIS X-ray contours showing the hard X-ray footpoint sources reported for a flare on 1980 May 21. The left panel shows contours in the 16–30 keV energy band and the right panel shows the same contours overlaid on an Hα image taken at nearly the same time. After [93]; © AAS. Reproduced with permission.

and a central fine field of view of 2.67 arc minutes, containing 8″ pixels. HXIS was the first to show double hard X-ray sources, as shown in Fig. 1.3. The hard X-ray fluxes from these two sources peaked within 3 s of each other, and both sources shared a location with an Hα flare kernel [92]. This indicated that the hard X-rays were most probably from accelerated electrons impacting the high density material at the footpoints of a magnetic loop, as indicated in the now classic model of a solar flare in Fig. 1.1.

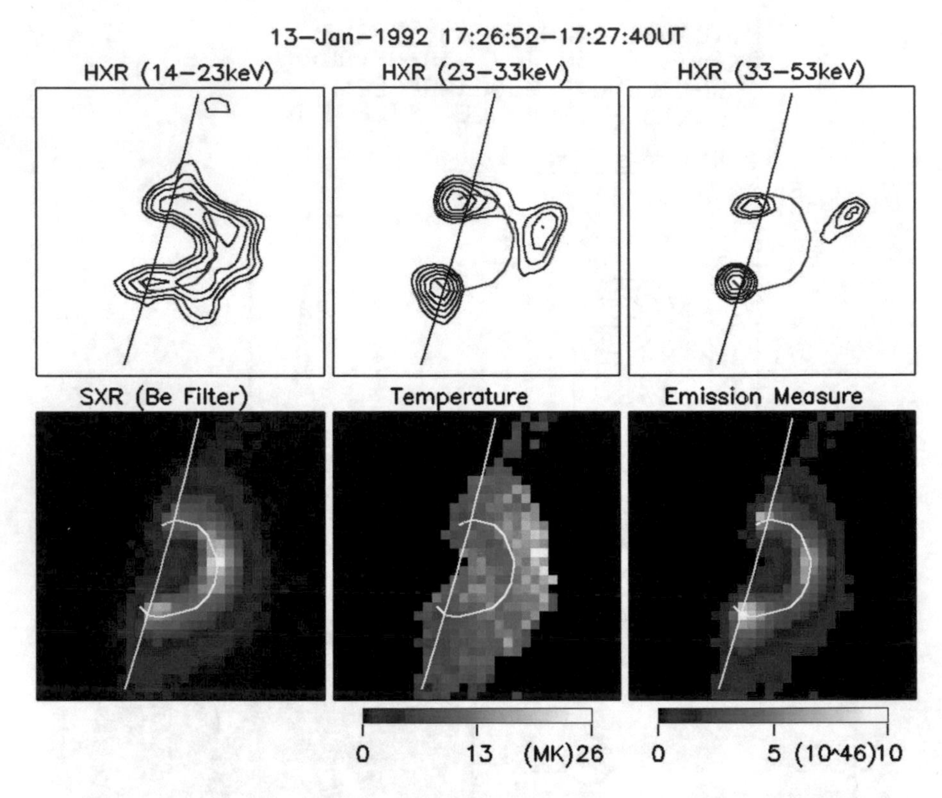

Fig. 1.4 Yohkoh X-ray images showing the "above-the-loop-top" hard X-ray source reported by [134] for a flare on 1992 January 13 near the west limb indicated by the diagonal solid black line in each image. The top row of images made from HXT observations in three energy ranges shows the bright hard X-ray footpoints on the solar disc with a third source seen most clearly in the two higher energy images well above the limb. This source is clearly at high altitude above the hot thermal loops seen in the soft X-ray images made with Yohkoh/SXT shown in the bottom row. After [134], used with permission of Springer.

HXT on Yohkoh used 32 pairs of multiple modulation collimators on a three-axis pointed spacecraft to measure both the sine and cosine parts of 32 spatial Fourier components (see Sect. 3.1). It covered the energy range from 14 to 100 keV in four broad bands and had $5''$ angular resolution and 0.5 s temporal resolution. It is perhaps best known for identifying [134] a hard X-ray source above the magnetic loops seen in soft X-rays (Fig. 1.4), thus providing strong evidence that the electrons producing the bremsstrahlung hard X-rays were accelerated as the result of magnetic reconnection in a coronal current sheet, as indicated in the flare model shown in Fig. 1.1.

RHESSI [125] is described in detail in Chap. 3 of this book, with many examples of its imaging capability noted in subsequent chapters. The example shown in Fig. 1.5 is included here for comparison with the images from the other three earlier instruments, primarily to illustrate the very significant evolution of instrument

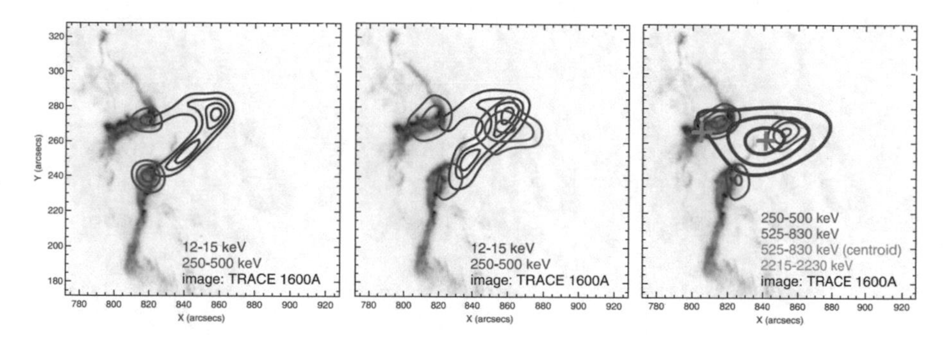

Fig. 1.5 *RHESSI* X-ray image contours overlaid on a UV image made with the Transition Region and Coronal Explorer (TRACE) instrument at 1600 Å for a flare on 2005 January 20. The 12–15 keV contours show the location of a hot thermal loop while the 250–500 keV contours show bright bremsstrahlung X-ray emission from electrons streaming down the legs of the loop to the two footpoints. In the central and right-hand frames, a third 250–500 keV source is evident nearly co-spatial with the top of the 12–15 keV loop. In the right-hand frame, this high energy source is also evident in the 525–830 keV energy band, although in that case only the centroid of the source could be determined, not its extent. After [117],© AAS. Reproduced with permission.

capabilities over the last four decades. These images show the expected bright footpoints now seen in most hard X-ray flares but also a strong coronal source extending to energies above 500 keV. In other flares, e.g., [179], a second coronal source is also seen at even higher altitudes suggesting that magnetic reconnection is taking place between them with electrons accelerated in both the upward direction away from the Sun and downwards towards the footpoints, as indicated in the current flare model shown in Fig. 1.1. Note that the centroid of emission in the 2215–2230 keV band shown in Fig. 1.5 was found to be nearly co-spatial with the northern footpoint. This energy band encompasses emission in the nuclear de-excitation line at 2223 keV that results from the capture of neutrons produced by accelerated protons and heavier ions. Thus, *RHESSI* provided the first evidence for possible differences in the location of accelerated electrons and ions in solar flares. This difference was further supported by *RHESSI* imaging observations [98] of earlier gamma-ray flares that occurred in October and November, 2003.

Finally, on February 2020 a new hard X-ray space telescope was launched by the European Space Agency (ESA) as part of the payload of the *Solar Orbiter* mission. The *Spectrometer/Telescope for Imaging X-rays* (*STIX*) has been designed according to an imaging concept that has several similarities with *RHESSI*, as will be discussed in Chap. 3.

Chapter 2
X-Ray Imaging Methods

Abstract Imaging in X-rays is inherently challenging because X-ray wavelengths are comparable to atomic dimensions (or, equivalently, X-ray photon energies are comparable to the binding energies of electrons in atoms), so that a ray-based focusing optics approach is not generally tenable. This problem becomes more acute at the shorter wavelengths corresponding to hard X-ray energies. In this chapter various X-ray imaging methods are described, from the basic medical technique using a single point source of X-rays to image body parts, to the techniques used in solar physics and astrophysics to image the X-ray sources themselves. X-rays interact with matter through both their wave and particle properties; the principal processes involved are absorption, scattering, reflection, refraction, and diffraction. Techniques are described that use these different forms of the interaction of X-rays with matter to obtain images of sources of astrophysical interest. Examples are provided of X-ray telescope designs to achieve the highest sensitivity, angular resolution, and dynamic range over hard X-ray energies ranging from 10 keV to 1 MeV.

2.1 Medical Imaging

The simplest (and earliest) X-ray imaging technique is radiography, widely used in the medical profession to make internal body parts visible, as illustrated in Fig. 2.1. It relies on the different X-ray absorption coefficient of different body parts, most notably bones and soft tissue. Using an essentially point source of X-rays, photons travel in straight lines between the source and an X-ray sensitive film or other recording device. Any object with a sufficiently high absorption coefficient that is placed between the source and the film will leave a dark (unexposed) pattern on the film, revealing the degree of X-ray absorption along the different paths through the object.

M. Piana et al., *Hard X-Ray Imaging of Solar Flares*,
https://doi.org/10.1007/978-3-030-87277-9_2

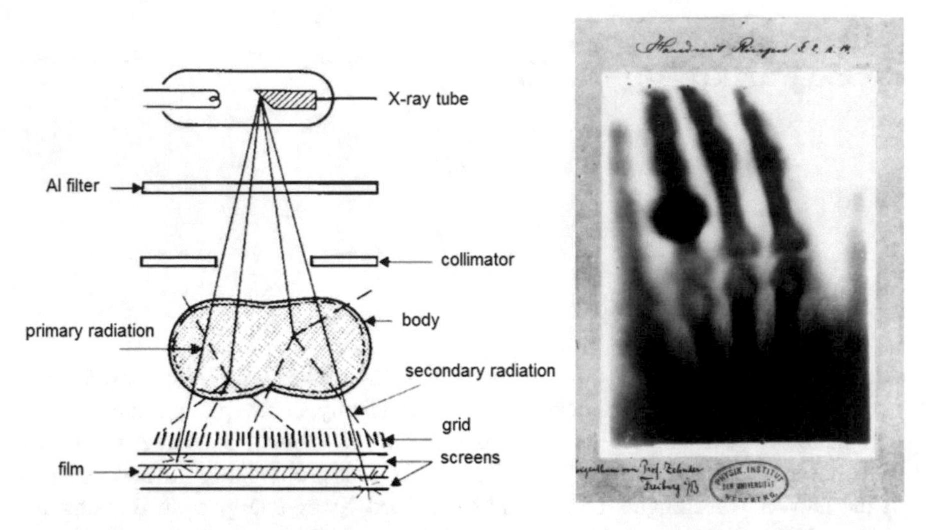

Fig. 2.1 Simplified layout of medical X-ray imaging. A single point source of X-rays is generated in an X-ray tube, typically by firing a focused electron beam at a tungsten target. The primary X-rays from this source propagate downwards and pass through the object of interest before interacting with the atoms of the film. Some of the X-rays are absorbed, depending on the density and composition of the material along the path through the object. An image of the object is obtained when the film is developed, with the dark areas corresponding to trajectories with the greatest absorption. After [200]; used with permission. The image on the right is of Röntgen's wife's hand.

Although various collimators, anti-scatter grids, and other more refined technologies have been developed to both improve the image quality and reduce the dose to the patient, the basic concept of radiography has remained the same since Wilhelm Röntgen first made the image of his wife's hand[1] (shown in Fig. 2.1) after discovering X-rays in 1895. Interestingly, modern focused anti-scatter collimators with narrow holes that diverge from a point X-ray source [192] were initially fabricated by Mikro Systems in Charlottesville, VA, using the same technique used to make the fine *RHESSI* grids (corresponding to detectors #1–4) for imaging of X-ray sources on the Sun (see Sect. 3.2).

Three-dimensional images can be built up using *computed tomography*, in which the X-ray source and detector are co-rotated about the object of interest as the object is moved slowly perpendicular to the beam plane creating a spiral pattern [144]. In this technique, the light absorption is related to the line integral of the tissue density and, therefore, the image reconstruction process relies on the inversion of the *Radon transform* [143].

[1] https://commons.wikimedia.org/w/index.php?curid=5059748.

2.2 Solar and Astrophysical X-Ray Imaging

The first X-ray telescope in astronomy, a pinhole camera on an Aerobee-Hi rocket, was used to observe the Sun on 1960 April 19 [21]. Since then, tremendous advances have been made in X-ray imaging of both solar and non-solar sources. Although these two fields cover similar X-ray production mechanisms and share similar imaging techniques, there are notable differences in design philosophy, driven by the vastly different distances to the respective sources.

The main requirement for detecting the relatively low photon fluxes from even the strongest astrophysical (cosmic) sources is high sensitivity, which means both a large sensitive area and a low background. Since most cosmic sources are so far away, there is no possibility of resolving spatial features. Thus, angular resolution is generally less important, although arcsecond angular resolution has been achieved at lower X-ray energies to resolve features in more extended sources such as supernova remnants [13].

Telescopes optimized for astrophysical observations are generally not suitable for solar observations. They are too sensitive for all but the quiet Sun, non-flaring active regions, or the weakest flares, and thermal constraints often prevent these astrophysical telescopes from even pointing at the Sun at all. Moreover, for solar observations, X-ray fluxes from the largest flares are vastly higher than for any cosmic source. Thus, the instruments have to be able to cope with such high fluxes without saturation, while still maintaining the ability to detect weak flares and the weaker fluxes at very high photon energies. This results in a required dynamic range of over 5 orders of magnitude. In addition, to be able to resolve physically interesting features on the Sun, angular resolutions of significantly better than $\sim 10''$ are required. Different available imaging techniques have been reviewed in [159].

2.3 X-Ray Imaging Techniques

There exist several X-ray imaging techniques that use different forms of the interaction of X-rays with matter to obtain images of sources of solar and astrophysical interest. Examples are described here of X-ray telescope designs using these different processes to achieve the highest sensitivity, angular resolution, and dynamic range over hard X-ray energies ranging from 10 keV to 1 MeV.

2.3.1 Absorption

The use of collimators, masks, and grids to record imaging information has been reviewed in [94]. Here we summarize the salient features of each method that uses X-ray absorption as the basic process for obtaining imaging information.

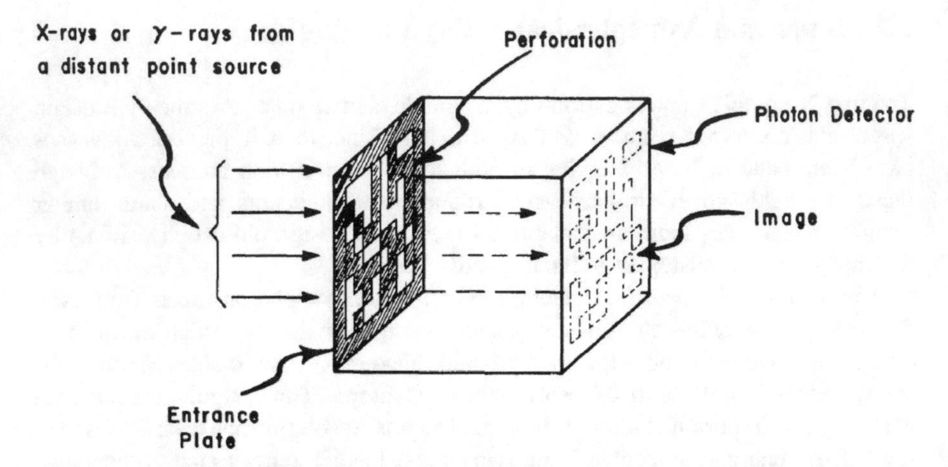

Fig. 2.2 Principle of X-ray imaging using a coded aperture mask. After [53], © AAS. Reproduced with permission.

2.3.1.1 Collimation

Imaging can be achieved using multiple narrow tubes with their long axes oriented towards different points on the source of interest. X-rays from any given point can reach a position-sensitive detector through only one of the tubes, and are detected with sufficient spatial resolution to identify the responsible tube. This was the operational principle of HXIS on SMM [194]. It used multiple holes in ten separated tungsten grid plates to form collimation paths with an angular resolution as fine as $8'' \times 8''$ FWHM over a $160''$ field of view. The position sensitive detector system consisted of 450 mini-proportional counters, each sensitive to the energy range from 3.5 to 30 keV.

2.3.1.2 Coded-Aperture Techniques

The pinhole camera used to make the first solar X-ray images absorbed incident X-rays from all directions except those that passed trough the pinhole [21]. An extension of that technique uses coded aperture masks (Fig. 2.2), essentially absorbing plates with multiple small holes in a coded pattern that casts its shadow on a position-sensitive detector [40]. Compact sources at different angular locations in front of the mask cast overlapping shadow patterns, requiring some form of reconstruction technique to determine the source locations from the pattern of photons counted in the detector plane. This reconstruction problem has been alleviated by the use of so-called Uniformly Redundant Arrays (URAs). The coded-mask technique allows for a significant increase in detector sensitive area while maintaining fine angular resolution over a wide field of view.

An example of a telescope that uses this technique is the Burst Alert Telescope (BAT) on the Neil Gehrels Swift mission, launched on 2004 November 20 [76]. This instrument is designed to detect and locate the source of 15–150 keV gamma-ray bursts over a wide field of view of 1.4 steradians, with an angular accuracy of 4 arcminutes. It uses a coded mask with a completely random pattern (with 50% open area and 50% closed) rather than the commonly used URA pattern.

2.3.1.3 Modulation-Collimator Techniques

A variation on coded-aperture techniques, aimed at achieving high angular resolution without the need for detectors with high spatial resolution, is the Oda collimator [145, 146]. The principle of operation is illustrated in Fig. 2.3, and is discussed in quantitative detail in Sect. 3.2. Two identical grids made of closely spaced wires or slats are spaced a distance D apart. This layout of slits and slats constrains possible arrival directions of photons that pass through both grids, thus providing angular resolution to within a few arc-seconds, depending on the widths (d) of the slats and the separation (D) between the grids. For sources with a FWHM angular size that is much smaller than d/D radians, the fraction of the intensity that is transmitted through both grids varies between two extremes, depending on whether the shadow of the slats of the front grid falls on the slats of the rear grid or on the gaps between them. Possible directions to the source can be determined by measuring

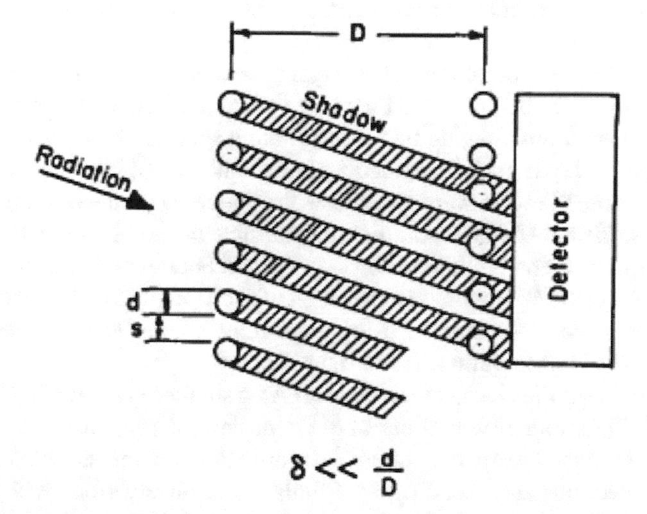

Fig. 2.3 Principle of operation of an Oda collimator, showing the shadow pattern from a front grid impingent on an identical rear grid. The intensity recorded in the detector depends on the direction to the source relative to the direction normal to the grids. See text for details. After [145], used with permission of OSA Publishing.

the transmitted flux as the orientation of the collimator axis is changed in a known way.

For more extended sources, the intensity modulation amplitude decreases with source size, basically because the different parts of the source modulate out of phase with each other. In practice, the modulation amplitude falls from 90% to 10% of its point-source value for Gaussian sources with FWHM $\simeq 0.5d/D$ and $\simeq 1.5d/D$, respectively. Thus, multiple collimators with different values of d/D must be used to cover a range of source sizes. For *RHESSI*, 9 different grid pitches were used with each d/D value increased by a factor of $\sqrt{3}$ to give a geometric series of angular resolutions ranging from $\sim 2''$ to $\sim 180''$ [96].

This basic principle was used originally for X-ray astronomy but has been used more recently for many solar X-ray imaging telescopes where fine angular resolution is more important than high sensitivity. Examples include the Solar X-ray Telescope (SXT) on the Japanese Hinotori satellite, launched on 1981 February 21, and *RHESSI* (see Sect. 3.1), launched on 2002 February 5. In both cases, the modulation of the transmitted flux was achieved by orienting the collimator axes close to the spacecraft spin axis and rotating the whole spacecraft about the direction to the Sun. In this way, detector counting rates are rapidly modulated as the spacecraft rotates, with the amplitude and phase dependent on the source location on the Sun and its angular extent. HXT on the Japanese Yohkoh mission [113] used a different technique to obtain directional information on this (non-spinning) spacecraft. In that case, two pairs of identical grids were needed to measure a single spatial Fourier component of the source, one to measure the sine component and the other (with its top and bottom grids shifted by 90° in phase) to measure the cosine component.

The Spectrometer/Telescope for Imaging X-rays (*STIX*) [120] on the Solar Orbiter mission, launched on 2020 February 9, uses a different technique to achieve high angular resolution imaging on a non-spinning spacecraft (see Sect. 3.6). In this case, the front and rear grids have identical pitch but the slits are oriented at a small angle to one another. In this way, a coarse Moiré pattern is produced on the detectors, which have sufficient spatial resolution to determine the amplitude and phase of the Moiré pattern. Each grid pair thus allows one spatial Fourier component of the image to be determined. Thirty-two such pairs, each with different pitches and orientations, allow images to be reconstructed using algorithms similar to those developed for analyzing *RHESSI* observations (see Chaps. 5 and 6).

The Micro Solar Flare Apparatus (MiSolFA), a smaller version of *STIX*, has been proposed [122] for joint observations of the same flares to provide three-dimensional stereoscopic views, which can provide information on the anisotropy of flare-accelerated electrons (see Sect. 1.2.2). Similar joint observations will be possible with the Hard X-ray Imager (HXI) on the Chinese Advanced Space-based Solar Observatory (ASO-S) [75] and [208]. HXI uses a similar technique to that used for HXT on Yohkoh but with Cerium-activated Lanthanum Bromide ($LaBr_3$:Ce) scintillation detectors to allow it to extend the high energy response up to 300 keV.

2.3.2 *Scattering*

Techniques using Compton scattering are limited to energies above ~100 keV and have relatively poor angular resolution. Thus, they have been used to locate astrophysical sources in the sky but not for solar X-ray imaging. The most notable instrument to use this technique was the COMPton TELescope (COMPTEL) on the Compton Gamma-Ray Observatory [166, 181] that operated from 1991 to 2000. The principle of measurement depended on the detection of the Compton scattered electron in an array of low-Z liquid scintillators and the subsequent detection of the scattered photon in a second array of high-Z, NaI(Tl) scintillators some 1.5 m away. The measured locations and energy losses in the two scintillator arrays allow the energy and arrival direction of the incident photon to be determined, the latter being along the edge of a cone whose axis is the direction of the scattered photon. COMPTEL covered energies from 750 keV to 30 MeV with an angular resolution of $\sim 1^\circ$, and performed a full-sky survey that is still unique in this energy range. Although no follow-up space mission has yet been approved, new telescopes using this technique have been built and flown on balloons, e.g., the Compton Spectrometer and Imager (COSI) [177] that can operate down to energies as low as ~100 keV.

2.3.3 *Reflection*

Because the wavelength λ of an X-ray is small compared to the separation a between the atoms or molecules of the material, X-rays at normal incidence to any material tend to penetrate into the material instead of being reflected. Thus, the reflectance (the fraction of the incident emission that is reflected) decreases very rapidly with decreasing wavelength, i.e., with increasing photon energy ϵ. At photon energies ϵ above ~100 eV, i.e., at EUV energies, the reflectance falls off[2] as ϵ^{-4}.

Useful reflectance can, however, be achieved at *grazing incidence*, which reduces the projected separation of the scattering centers and so effectively increases the ratio λ/a. By analogy with stones skipping along the surface of water, or the bouncing bombs dropped during the famous "dam busters" operation during the Second World War, a much greater fraction of the incident radiation is reflected at grazing incidence than for normal incidence. Such grazing incidence optics have been successfully and extensively used at soft X-ray wavelengths, but the technique requires extremely smooth surfaces. Bringing an X-ray beam to a sharp focus using grazing incidence reflection is generally achieved using a double reflection with the Wolter-1 configuration[3] shown in Fig. 2.4 [203–205]. The first reflection is off

[2] http://www.rxollc.com/technology/index.html.

[3] https://en.wikipedia.org/wiki/Wolter_telescope.

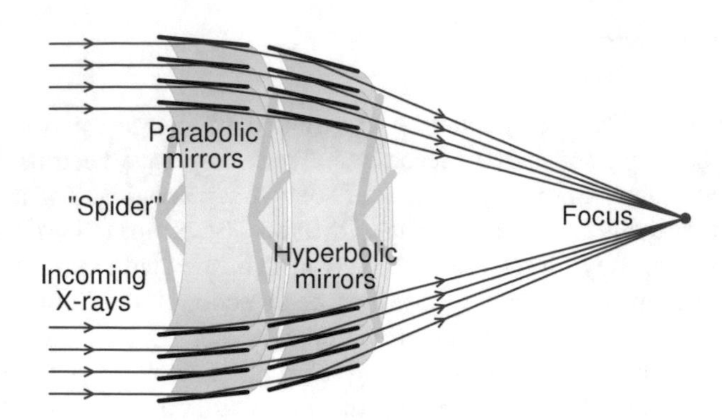

Fig. 2.4 Principle of a Wolter1 X-ray telescope design. Focusing is achieved through two grazing-incidence reflections, the first on a parabolic mirror and the second on a hyperbolic mirror. Multiple nested mirrors are used to increase the total collecting area. (By C. M. G. Lee—Own work, CC BY-SA 4.0, https://commons.wikimedia.org/w/index.php?curid=82992872).

a parabolic surface and the second is off a hyperbolic surface. Multiple shells are incorporated to increase the total collecting area.

Such a configuration is used by the NuSTAR mission [85] that was launched on 2012 June 13, becoming the first focusing high-energy X-ray telescope in orbit. It has a total of 133 nested shells, each with the two cylindrically-shaped segments required by the Wolter-1 design to bring X-rays from infinity to a focus at 10.15 m. All shells are coated with depth-graded multilayers, the inner 89 using Pt/C and the outer 44 using W/Si. It operates between 3 and 79 keV with an angular resolution of 18″ (FWHM), extending the sensitivity of focusing far beyond the $\sim$10 keV high-energy cutoff achieved by all previous X-ray satellites.

Much greater reflectance can be achieved at higher energies by using multilayer coatings that utilize coherent Bragg reflection, i.e., constructive interference between the beams reflected from the different layers. If the layers are all the same thickness, then the reflectance is sharply peaked at a specific photon energy. Such normal-incidence multilayer telescopes have been used extensively for narrow-band EUV imaging, e.g., the Atmospheric Imaging Assembly (AIA) on-board the Solar Dynamics Observatory (SDO) [124], the High Resolution Coronal Imager (Hi-C) [104], and the Solar Ultraviolet Imager (SUVI) on the Geostationary Operational Environmental Satellites-R Series spacecraft (GOES-R/SUVI) [129]. For broad-band reflectance extending into the hard X-ray energy range, depth-graded multilayers are used. In this case, the layer thickness is decreased with depth according to a power-law function, for example, allowing a broad energy range to be covered. Examples of the reflectance vs. photon energy that can be achieved for four different materials are shown in Fig. 2.5.

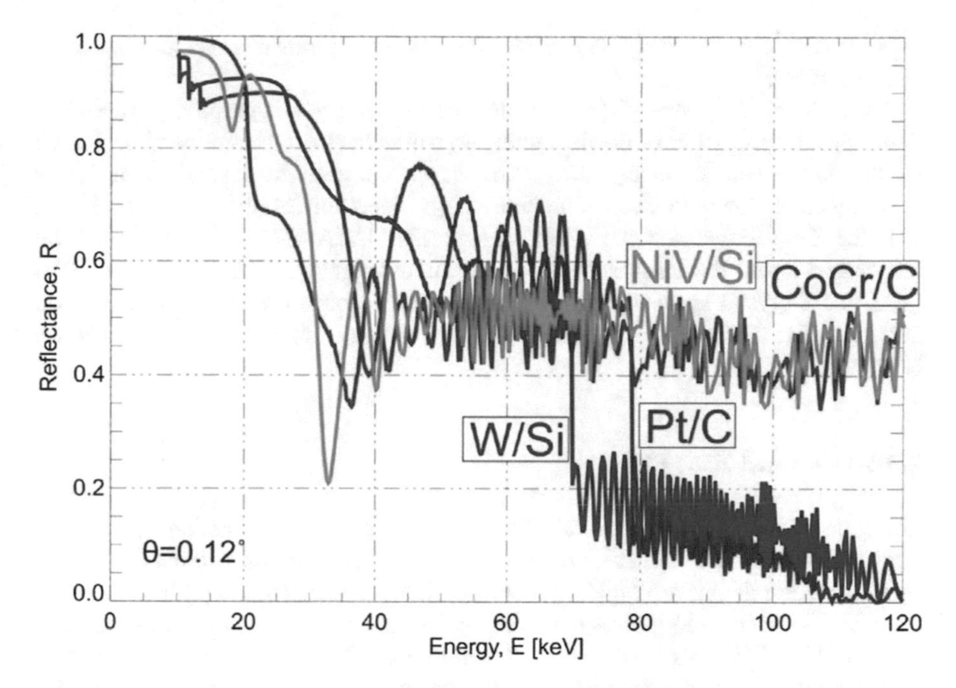

Fig. 2.5 Theoretical reflectance vs. X-ray photon energy for grazing incidence optics with depth-graded multilayers of various materials. Note the rapid falloff at the W K-edge at 70 keV and the Pt K-edge at 78 keV, limiting the useful energy ranges for these materials. Ni- and Co-based multilayers offer the promise of continuous energy response to higher energies. After [202], used with permission from SPIE.

2.3.4 *Diffraction*

2.3.4.1 Bragg Diffraction

As discussed above, X-ray photons tend to pass through or into most materials rather than being reflected at the surface. However, materials with a regular crystalline structure do reflect X-rays preferentially at the Bragg angle that is dependent on the X-ray wavelength (λ) and the distance (d) between the planes of atoms in the crystal. Reflection occurs when X-rays are incident at the surface of the crystal at an angle θ to the crystal surface[4] given by

$$2d \sin\theta = n\lambda\,, \tag{2.1}$$

where n is any positive integer. At that angle, the path difference between the beams reflected at adjacent crystal layers is an integer number of wavelengths and so the

[4] Note that θ is *not* the angle to the normal, as used, e.g., in Snell's Law.

rays interfere constructively. A similar effect occurs in transmission and is known as Laue diffraction.

Both Bragg and Laue diffraction are used in X-ray topography, primarily to determine the properties of the diffracting crystal rather than the angular distribution of the X-ray source. Bragg diffraction has been used in several solar X-ray spectrometers flown to determine the energy spectrum of the incident X-rays, e.g., the X-ray Polychrometer (XRP; [1]) on the NASA SMM, launched on 1980 February 14, and the Rentgenovsky Spekrometr Izognutymi Kristalamy (RESIK; [183]) bent crystal spectrometer on the Russian Coronas-F mission, launched on 2001 July 31. At the time of writing, no X-ray imager using Bragg diffraction has been flown.

2.3.4.2 Fresnel Zone Plates

Since path-length differences of order one wavelength are needed to produce constructive interference, the short wavelengths of X-rays and gamma-rays mean that diffraction angles are typically very small. Nevertheless, X-ray imaging based on diffraction can be implemented using variations on the basic Fresnel zone plate concept [174, 175]. In such a design, shown in Fig. 2.6a, multiple concentric circles of alternately X-ray opaque and transparent materials are arranged so that first-order diffraction brings X-rays of a specific wavelength from a point source on one side of the plate to a focal point on the other side. The zones are dimensioned such that X-rays passing through adjacent open zones will have a path length difference of just one wavelength and hence interfere constructively at the focal point. In Fig. 2.6b, the shaded zones are not opaque but transmit radiation with a phase shift of π and so will also interfere constructively at the focal point. In Fig. 2.6c, the thickness of the shaded zones is everywhere such that the phase is shifted by the optimum angle to contribute the maximum flux at the focus and hence gives the highest possible effective area for such a device.

An alternative concept is the *photon sieve*, which has a series of pinholes arranged in a pattern similar to the rings in a Fresnel zone plate (e.g., [103]). These have the advantage of ease of fabrication and the ability to achieve a sharper focus but they do not make use of the full available area of the device. Consequently, they tend to be used for longer wavelengths, where source fluxes are higher and counting statistics is not as significant a problem as in X-ray and gamma-ray astronomy.

To date, no X-ray or gamma-ray imaging telescopes have been built and flown using diffraction optics. However, the prospect of such superb angular resolution possibilities that are perhaps six orders of magnitude better than the current state of the art means that they will eventually be implemented. The major problem is the long focal lengths needed to achieve the desired angular resolution. For X-ray imaging in the complex of iron lines at 6.7 keV for example, a focal length of $\sim$100 m is required to achieve better than $0.1''$ angular resolution [51]. Achieving the required micro-arcsecond resolution to image the accretion disk around a black hole would need a lens with a focal length of hundreds or even thousands of km and

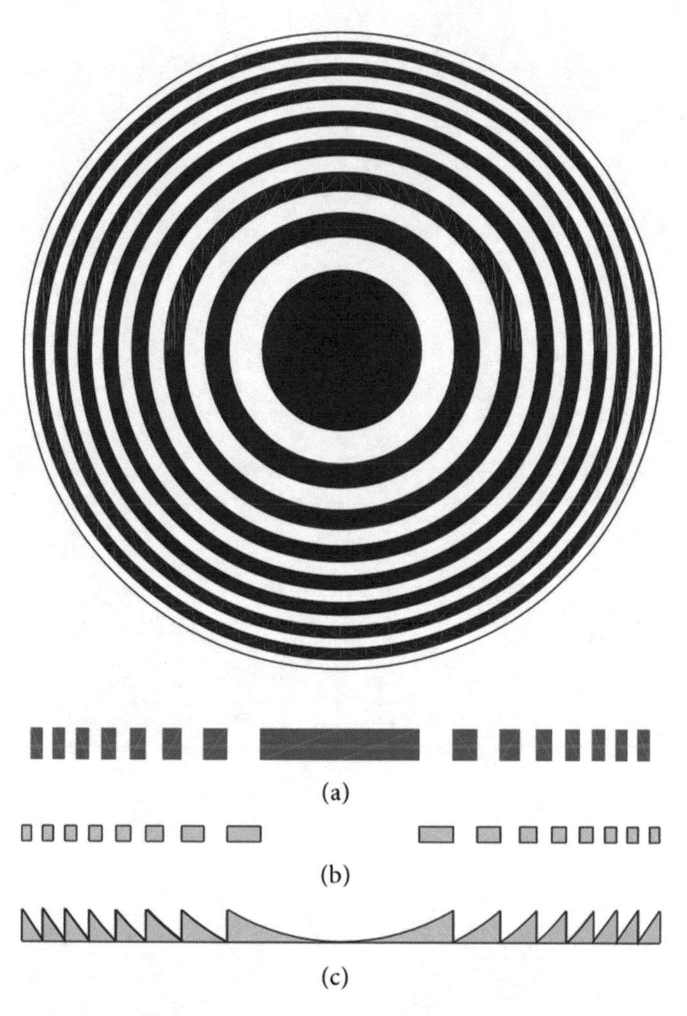

Fig. 2.6 Three types of diffractive optics for X-ray imaging. The basic Fresnel zone plate design is shown at the top of the figure and in cross section below in (**a**). In this case, the zones are alternately completely X-ray opaque and completely transparent. In (**b**), the shaded zones are not opaque but transmit radiation with a phase shift of π. In (**c**), the thickness of the shaded zones is everywhere such that the phase is shifted by the optimum angle. The profile in (**c**) is drawn for a converging lens assuming that the refractive index of the material is less than unity. After [175], used with permission from Hindawi.

would require the development of precise formation flying capabilities over such vast distances [176].

Chapter 3
RHESSI and *STIX*

Abstract In this chapter we describe the design features of the *RHESSI* and *STIX* instruments, both of which use bi-grid modulation collimators pointed at the Sun to obtain information that can be used to make X-ray images. The difference is that *RHESSI* is on a rapidly rotating spacecraft while *STIX* is on a 3-axis stabilized spacecraft. We describe the manner in which imaging information is encoded by the bi-grid collimators and the strengths and limitations of the two imaging concepts necessitated by the different spacecraft motions. We also describe the software that has been developed to transform the raw data collected into higher-level products useful for scientific studies. We illustrate the power of such imaging techniques using an example solar flare.

3.1 *RHESSI* Design/Brief History of Concept Development

RHESSI [125] was designed with the primary scientific objective of understanding particle acceleration and explosive energy release in magnetized plasmas at the Sun. Previous X-ray measurements, coupled with measurements at other wavelengths, had shown that in order to achieve that goal, hard X-ray imaging was required with an angular resolution in the few-arc-second range, at energies between $\sim$10 keV and 100 keV. Such an angular resolution was necessary to isolate the chromospheric footpoint sources (Fig. 1.1) and distinguish them from the hot coronal sources, and to determine the spectra of the bremsstrahlung emission from nonthermal electrons at energies well above that of the thermal emission from plasma with the highest temperatures thought to be $\sim$50 MK. In addition, it would be necessary to have an effective sensitive area of $\sim$100 cm^2 in order to detect weaker events, but the instrument would also have to be designed to avoid saturation for the largest events. This corresponds to a requirement for handling a dynamic range of some 10^5 in flare X-ray intensity from small GOES[1] B-class events to the largest X10-class events.

[1] Solar flares can be classified in many ways. One of the most common uses the observed flux in the 0.1–0.8 nm "soft" X-ray range, as observed by the series of Geostationary Operational

M. Piana et al., *Hard X-Ray Imaging of Solar Flares*,
https://doi.org/10.1007/978-3-030-87277-9_3

After reviewing the different available techniques for imaging hard X-rays (discussed in Chap. 2), it was realized that the only practical method of satisfying these observational requirements within the constraints of NASA's SMEX program was to use collimator-based Fourier-transform imaging [95] as described in Sect. 2.3.1.3 and, with only a single instrument involved (shown in Fig. 3.1), to simply rotate the whole spacecraft in a spin-stabilized configuration. The use of rotating modulation collimators (RMCs) allowed *RHESSI* to measure many (over a hundred) spatial frequency (i.e., Fourier) components, covering a broad range of angular size scales from $2 - 180''$. This compared favorably to the few tens of Fourier components that had been possible with previous imagers on 3-axis stabilized spacecraft (HXT on Yohkoh, for example, measured just 32 components covering spatial scales from $8 - 20''$).

Another advantage of using rotating modulation collimators is that the detectors need no spatial resolution—all the imaging information is encoded in the temporal modulation of the total X-ray flux passing through each collimator as the spacecraft rotates. This allowed a single large-area hyperpure germanium detector [178] to be used behind each collimator. Each detector was segmented into a front segment, optimized for lower energy photons, and a rear segment, optimized for higher energy photons up to the gamma-ray range. The fine ($\sim$1 keV FWHM) energy resolution of these detectors allowed accurate measurement of the exponentially falling thermal bremsstrahlung spectra (cf. Eq. (1.2)) and even power-law nonthermal spectra that can be as flat as ϵ^{-2} or as steep as ϵ^{-10}. An added bonus of these detectors is their large electrically-segmented volumes, allowing sensitive measurements of spectra from $\sim$3 keV to $\sim$3 MeV in the front segments and from $\sim$100 keV to $\sim$17 MeV in the rear segments. Note that photons scattered off the Earth's atmosphere can enter the detectors from the side and be recorded at lower energies, down to the $\sim$20 keV electronic threshold for the rear segments.

3.2 The *RHESSI* Imaging Concept

The basic *RHESSI* imaging concept is described in detail in [96]. The *RHESSI* instrument (shown in Fig. 3.1) encodes imaging information through the RMC technique described in Sect. 2.3.1.3, in which a set of parallel occulting grids rotates with the spacecraft, sweeping possible observing directions across the solar disk and

Environmental Satellites (GOES). Similar to the Richter scale for earthquakes, the classification scale is logarithmic: A-class events have fluxes less than 10^{-7} Watts per square meter (W m^{-2}), B-class events have fluxes between 10^{-7} and 10^{-6} W m^{-2}, C-class events have fluxes between 10^{-6} and 10^{-5} W m^{-2}, M-class events have fluxes between 10^{-5} and 10^{-4} W m^{-2}, and X-class events have fluxes exceeding 10^{-4} W m^{-2}. Within each class, a number indicates the multiplier of the lowest flux appropriate to that classification; for example, an $M5$ event has a 0.1–0.8 nm flux of 5×10^{-5} W m^{-2}. The most intense flare recorded by GOES in the *RHESSI* era was an X28 event (with an 0.1–0.8 nm flux of $28 \times 10^{-4} = 2.8 \times 10^{-3}$ W m^{-2}) that occurred on 2003 November 4.

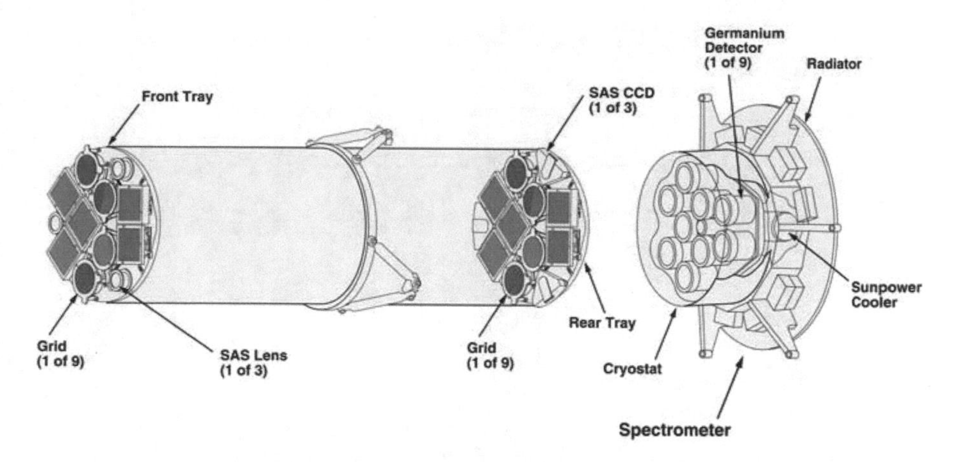

Fig. 3.1 Perspective of *RHESSI* showing the components necessary for imaging. Two essentially identical sets of nine grids are mounted on front and rear grid trays, with all the grids made of tungsten except for the finest grid pair that was made of molybdenum for ease of etching. A corresponding set of nine cooled germanium detectors is mounted behind the rear grids. The solar aspect system (SAS) consists of three lenses mounted on the front grid tray, which focus optical images of the Sun onto three linear diode arrays (labeled SAS CCDs in the figure) on the rear grid tray. It provides sub-arcsec knowledge of the radial pointing with respect to Sun center. Two optical roll angle systems (RASs, not shown) pointed perpendicular to the spin axis detect multiple stars each spacecraft rotation and provide the necessary roll angle information. The combined SAS and RAS data enables the absolute orientation of the grids to be determined on millisecond time scales and allows X-rays sources to be located on the solar disk to sub-arcsec accuracy. After [96], used with permission from Springer.

rapidly modulating the detected photon flux in the process. Imaging information is encoded [96] in the amplitudes and phases of the X-ray counting rate modulations in the different detectors as the spacecraft rotates at nominally 15 rpm. *RHESSI* is thus termed a "Fourier imager" because it provides imaging information through measurement of the spatial (or, more accurately, angular) Fourier components of the X-ray source(s) that produce these modulated time profiles.

To see how this works, consider one of the front grids defined by a number of parallel slits (apertures) and slats (X-ray opaque grid material), with the width of each slit and slat equal to $p/2$, so that the grid "pitch" (the spatial period) equals p and the transmission factor is 50%. The separation between this front grid and the corresponding rear grid with an identical slit/slat geometry is equal to L (see Fig. 3.2). Illumination of the front grid by parallel rays from a distant point source (distance $\gg L$) produces a periodic rectangular shadow pattern on the rear grid. When the bright strips produced by transmission through the front grid slits align with the opaque rear grid slats, no signal is observed by a detector situated below the rear grid. On the other hand, when the bright strips produced by transmission through the front grid slits align with the slits on the rear grid, a maximum in intensity is detected equal to one-half the brightness of the source. The transmitted intensity thus rises and falls (as we shall show shortly) approximately linearly with

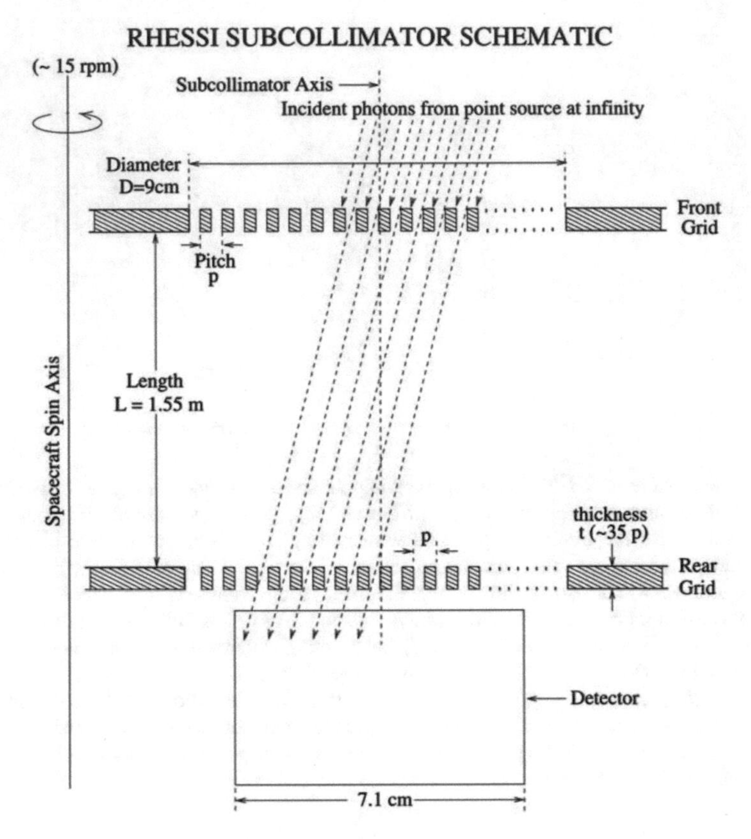

Fig. 3.2 Geometry of the *RHESSI* imaging process. Light from a distant point source passes through two identical grids, each with a slit/slat pitch of p and separated by a distance L. A detector records a photon only if it passes through both grids; thus the detected flux depends on the orientation of the grids relative to the source direction. In the spacecraft frame, this direction changes continually due to spacecraft rotation, thus providing a temporally modulated signal that provides information on the source direction. After [96], used with permission from Springer.

time, rising from a minimum of zero to a maximum of one-half the source intensity and then falling back down to zero again. The average detected flux is just one-fourth the source flux that would have been detected in the absence of the grids.

Now consider a point source, located at polar coordinates (θ_o, ϕ_o), that sends N photons s^{-1} into the spacecraft telescope aperture. (Here θ_o is the angle between the vector to the source and the spacecraft rotation axis, and ϕ_o is the azimuthal angle measured around this axis, measured from a reference direction (e.g., the direction parallel to the grid slits).) At any source location, the front grid allows $N/2$ photons s^{-1} to be transmitted through it. How many of those photons are in turn transmitted through the rear grid, and hence onto the detector, depends on the location of the front grid shadow pattern relative to the slit/slat pattern on the rear grid.

The temporal modulation of the signal is best understood by considering the view from a frame of reference fixed on the rotating spacecraft. As the spacecraft on which the grids are mounted rotates, an observer on the spacecraft sees the source appear to move along a circular path, centered on the spacecraft rotation axis and in the opposite sense to the spacecraft rotation. This circular motion has vector components both parallel and perpendicular to the slits in the occulting grids. The component of motion parallel to the slits results in very little change in the shadow pattern cast by the front grid and hence little change in the intensity seen by the detector. However, the component of motion perpendicular to the slits causes the shadow pattern produced by the front grid to move across the rear grid in a direction perpendicular to the slits [96]. This causes the fraction of the source intensity that is transmitted to vary with the perpendicular offset angle from a minimum of zero (when the dark rectangles of the front grid shadow pattern align with the slits in the rear grid) to a maximum of $N/2$ photons s^{-1} (when the dark rectangles of the front grid shadow pattern align with the slats in the rear grid).

This perpendicular motion, the projection of the apparent circular motion of the source on the sky onto the direction perpendicular to the slit orientation, is a simple harmonic oscillation with period, T, equal to the rotation period of the spacecraft (typically ∼4 s for RHESSI). The shadow pattern from the front grid is thus projected onto the rear grid at a time-dependent angle

$$\theta(t) = \theta_o \left| \sin\left(\frac{2\pi t}{T}\right) \right| \ , \tag{3.1}$$

where we have chosen the time origin to be when the source azimuthal angle $\phi_o = 0$, so that the shadow of the slats in the front grid fall exactly on the slats in the rear grid, corresponding to maximum transmitted intensity.

At $t = 0$, $\theta = 0$ and the X-ray photons traveling through the slits in the top grid pass directly through the slits in the bottom grid, corresponding to a maximum in intensity. As t and θ increase, a progressively greater fraction of each dark rectangle in the shadow pattern encroaches into the corresponding lower grid slits, and the intensity of X-rays passing through both grids onto the detector drops. Since θ is a very small angle (the field of view of *RHESSI* is of the order of a degree or so, and the maximum value of θ_o for a solar source is half a degree even for a rotation axis pointed at the solar limb), we can approximate $\tan\theta \simeq \theta$. Thus, at an angle $\theta(t)$, the front grid shadow pattern is displaced by a distance $x(t) \simeq L\theta(t)$, and so the number of photons per second transmitted through the rear grid onto the detector is

$$n(t) = \frac{N}{2}\left(1 - \frac{x(t)}{p/2}\right) = \frac{N}{2}\left(1 - \frac{L\,\theta(t)}{p/2}\right) \simeq \frac{N}{2}\left(1 - \frac{\theta_o \left|\sin\left(\frac{2\pi t}{T}\right)\right|}{p/2L}\right) . \tag{3.2}$$

This equation shows that the intensity decreases as a linear function of the offset angle $\theta(t)$ perpendicular to the slits, from a maximum of $N/2$ at $\theta = 0$ to zero when $\theta_o \sin(2\pi t/T) = p/2L$, i.e., when the dark rectangles in the shadow pattern from the front grid are coincident with the slits on the rear grid. Further drift of the front grid shadow pattern then causes the intensity to increase back to a maximum of $N/2$ when $\theta_o \sin(2\pi t/T) = p/L$, i.e., as the dark rectangles in the front grid shadow pattern move to locations coincident with the slats of the rear grid, one grid pitch removed from the shadow locations when $t = \theta = 0$. As the spacecraft continues to rotate, the intensity again decreases, and so on. The observed rate of photon detections $n(t)$ thus rises and falls in a cyclical saw-tooth pattern, from a minimum of zero to a maximum of $N/2$.

The number of intensity cycles, m, per spacecraft rotation period T depends on the value of the source polar angle θ_o: $m = 4\theta_o/(p/L)$. Sources located at greater polar angles θ_o have a greater range of offset angles θ (from 0 to θ_o) and so produce a higher number of intensity cycles per spacecraft rotation. Thus, rather straightforwardly, the *number* of intensity cycles per spacecraft rotation provides an immediate determination of the source polar angle θ_o. Determination of the source azimuthal coordinate ϕ_o follows from consideration of the *phase* of the intensity cycle pattern relative to the phase of the slit orientation, measured with respect to a reference direction on the sky (e.g., celestial North or, more usefully for solar observations, the northern direction on the solar disk.).

The above argument applies strictly to point sources. For extended sources, the intensity modulation pattern is a superposition of the intensity patterns for a set of many point sources, each with intensity $N(\theta_o, \phi_o)$. This is illustrated in Fig. 3.3, where each panel represents the response to different source sizes and locations.

- *Panel 1.* The reference modulation pattern produced by a point source with unit intensity, located at a polar angle θ_o (represented by the angular distance of the source from the rotation axis) and azimuthal angle $\phi_o = 0$. The intensity varies from near zero (when the slat shadow pattern of the top grid—containing half the intensity of the source—aligns with the slit pattern of the bottom grid), to a maximum of one half (when the slat shadow pattern of the top grid aligns with the slat pattern of the bottom grid, so that there is 100% transmission through the bottom grid). The measured intensity averaged over a full spacecraft rotation is thus one quarter of the source intensity while the amplitude of the modulated signal (peak to valley) is half the source intensity.
- *Panel 2.* The modulation pattern produced by a point source located at the same polar angle and azimuth, but with half the intensity.
- *Panel 3.* The modulation pattern produced by a point source with unit intensity, located at the same polar angle θ_o but at a different azimuthal angle $\phi_o = \pi/4$. The modulation pattern is the same as in Panel 1, but shifted in time by one-eighth of a cycle.
- *Panel 4.* The modulation pattern produced by a point source with unit intensity, located at azimuthal angle $\phi_o = 0$, but a larger polar angle θ_o. The modulation pattern has the same average intensity and modulation amplitude as in Panel 1, but has more cycles per spacecraft rotation period.

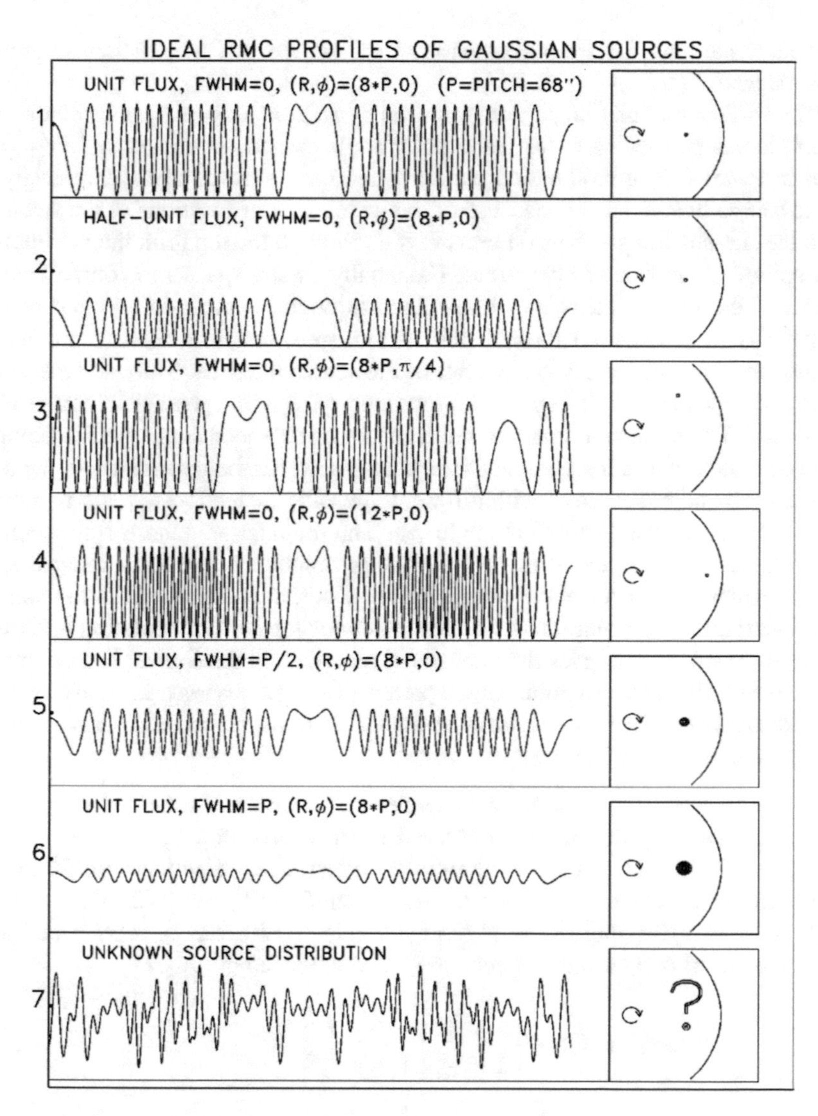

Fig. 3.3 Sample modulation patterns produced by sources of various sizes at various polar coordinates relative to the rotation axis of the spacecraft. Each of these profiles constitutes a "basis function" that can be used to deconstruct an observed modulation pattern (bottom panel) into a sum of point sources of different intensities, i.e., an image. After [96], used with permission from Springer.

- *Panel 5.* The modulation pattern produced by an extended source with unit intensity, located at a polar angle θ_o and azimuthal angle $\phi_o = 0$. The modulation pattern has the same average intensity at the source in Panel 1. However, different parts of the source create different locations of the shadows in the slat-shadow patterns on transmission through the top grid and, because of overlap of the shadow patterns from different parts of the source, the modulation amplitude is reduced.

- *Panel 6.* The modulation pattern produced by an even more extended source with unit intensity, located at the same polar angle θ_o and azimuthal angle $\phi_o = 0$ as in Panel 1. The modulation pattern again has the same average intensity as the source in Panel 1, but because of the wide range of locations of the shadows in the slat-shadow patterns on transmission through the top grid, the modulation amplitude is reduced to near-zero. Essentially, at any spacecraft rotation angle, half of the source intensity is transmitted through the top grid and half of that intensity is transmitted through the bottom grid, resulting in a near-constant observed intensity equal to a quarter of the source intensity. Thus, the measured intensity averaged over a spacecraft rotation period is independent of source size.
- *Panel 7.* The modulation pattern produced by an extended source with a complicated structure. Such a complex source can, however, be considered as the sum of a set of point sources i with different intensities, each located at a particular polar angle θ_{oi} and azimuthal angle ϕ_{oi}. The modulation pattern from such an extended source is the sum of the various point-source modulation patterns, each pattern being weighted by the intensity at that location within the source. Evaluating the contribution of each point-source ("basis") modulation pattern to the observed pattern gives the intensity at that point in the source; i.e., an image. Deconvolution of these point source patterns from an observed intensity vs. time modulation profile is the essence of the CLEAN image reconstruction method discussed in Sect. 5.2 below.

For $\theta_o \gg p/2L$ the number of cycles per spacecraft rotation period m is quite large, and the variation of observed intensity $n(t)$ throughout one cycle is approximately linear; $n(t)$ has a "saw-tooth" pattern. The triangular profile for the single cycle that extends over $-p/2L\theta_o < \sin(2\pi t/T) < p/2L\theta_o$, i.e., from $-\tau/2 < t < \tau/2$ (where $\tau = (1/2\pi)\,(1/\theta_o)\,(p/L)\,T$), can be represented as a Fourier series of odd cosine components (cf. equation (2) of [96]):

$$
\begin{aligned}
n(t) &= \frac{N}{2} \times \begin{cases} 1 + \frac{2t}{\tau} \; ; \; -\frac{\tau}{2} \le t \le 0 \\ 1 - \frac{2t}{\tau} \; ; \; 0 \le t \le \frac{\tau}{2} \end{cases} \\
&= \frac{N}{4} \left[1 + \frac{8}{\pi^2} \sum_{k=1,3,5,\cdots}^{\infty} \frac{1}{k^2} \cos \frac{2\pi k t}{\tau} \right] .
\end{aligned}
\tag{3.3}
$$

Although this expression[2] is strictly valid only for the central triangular "sawtooth" of the modulation pattern, in practice the number of cycles per spacecraft rotation is sufficiently large that Eq. (3.3) can be applied throughout the entire modulation

[2] The above analysis applies when the occulting grids have exactly equal slit and slat widths; then only odd cosine harmonics are present. For unequal slit/slat widths, the Fourier expansion includes even cosine harmonics as well.

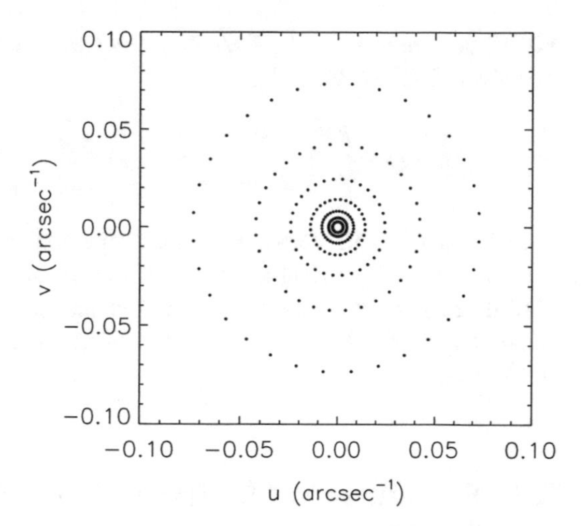

Fig. 3.4 Spatial frequency points sampled by the *RHESSI* instrument (here, only detectors #3 through 9 are represented). The sampled points lie on a series of circles, with radii corresponding to the reciprocals of the angular resolutions of the *RHESSI* grid pairs. After [57]; used with permission from EDP Sciences for European Southern Observatory.

profile. When $t = 0, \tau, 2\tau, \cdots$, all the cosine terms have value unity, the infinite sum is $\sum_{k=1,3,5,\dots}^{\infty} 1/k^2 = \pi^2/8$ and $n(t) = N/2$. On the other hand, when $t = \tau/2, 3\tau/2, \cdots$, all the cosine terms have value -1, the infinite sum is $-\sum_{k=1,3,5,\dots}^{\infty} 1/k^2 = -\pi^2/8$ and $n(t) = 0$.

Taking the lead ($k = 1$) term in the sum in Eq. (3.3) gives

$$n(t) \simeq \frac{N}{4}\left[1 + \frac{8}{\pi^2}\cos\frac{2\pi t}{\tau}\right], \tag{3.4}$$

from which we see that each detector essentially measures, through its modulation profile, the extent to which the source has structure on a time scale τ, corresponding to an angular scale $\alpha = p/L$ measured perpendicular to the orientation of the grid slit/slat array. Sources that have an angular extent $\Delta > p/L$ will produce a front-grid shadow pattern that is significantly reduced in contrast (see Panels 5 and 6 of Fig. 3.3); indeed, for sufficiently extended sources ($\Delta \gtrsim 3p/L$) the shadow pattern vanishes; exactly one-half of the source intensity is transmitted through the front grid as a uniformly bright pattern, and exactly half of this, or one-fourth of the original source intensity, is transmitted through the bottom grid to arrive at the detector. This lack of modulation means that such an extended source cannot be effectively imaged (or even located) by grids with $p/L < \Delta$.

To summarize, *each set of RMC grids effectively determines a set of spatial (or, more accurately, angular) Fourier components of the source, termed visibilities*[3] *([96, 97], see Sect. 6), all with the same magnitude in the angular frequency domain, and with an orientation that varies as the spacecraft and instrument co-rotate.* This angular frequency domain is denoted as (u, v) space, where u and v have units of arcsec^{-1}. Figure 3.4 shows the locations of the (u, v) points at which these source

[3] https://sprg.ssl.berkeley.edu/~tohban/wiki/index.php/RHESSI_Visibilities.

visibilities are sampled by *RHESSI*. The formal equation linking the image $I(x, y)$ and the visibilities $V(u, v)$ is

$$V(u, v) = \int_{x=-\infty}^{\infty} \int_{y=-\infty}^{\infty} I(x, y)\, e^{2\pi i [u(x-x_o)+v(y-y_o)]}\, dx\, dy\,, \tag{3.5}$$

where (x_o, y_o) is the location of the projection of the spacecraft rotation axis onto the solar disk. Obtaining the form of the image $I(x, y)$ from the set of measured visibilities $V(u, v)$, located at the limited and sparsely distributed set of (u, v) values illustrated in Fig. 3.4, is the objective of many of the procedures discussed in this book.

3.3 Strengths and Limitations of the *RHESSI* RMC Imaging Technique

The RMC design of *RHESSI* provides images in a way that is not as intuitive as the pixel-by-pixel scanning techniques used in, for example, the optical domain. Compared to such straightforward imaging, the RMC visibility-based method provides both advantages ("strengths") and disadvantages ("limitations").

Strengths

1. *Range of Source Sizes.* By using grid pairs of different pitches, the RMC method resolves X-ray sources on a range of physically meaningful size scales, from $\lesssim 2''$ to $\gtrsim 180''$. Because of the intrinsically distinct modulation patterns corresponding to sources of different spatial extents, it also allows sufficiently intense extended (e.g., coronal) sources to be detected even in the presence of strong compact (e.g., footpoint) sources.
2. *Detector Design Simplicity.* Since the detectors need only measure total flux, they do not need to have any imaging capability. (Note that all nine *RHESSI* detectors are electrically separated into two segments, front and rear, but none of the segments provide any spatial information about the interacting photons.) The simplicity of the detector design, in principle, allows images to be made over a wide energy range ($\sim$3 keV to 17 MeV) in both the X-ray and gamma-ray domain. Compton scattering in the grid material does limit the imaging capability in the gamma-ray range, except for the special case of the 2.223 MeV neutron capture line, where images have been successfully reconstructed for several large flares [98].
3. *Imaging Spectroscopy.* Combining energy resolution and imaging information allows true *imaging spectroscopy*, where an energy spectrum is obtained for each sub-region in a recovered image.
4. *Dynamic Range in Overall Source Brightness. RHESSI* covers a very wide dynamic range of flare intensities, without significant pulse pileup (see Limitation #2 below) or saturation. This is achieved largely by employing thin

aluminum attenuators that absorb many of the low-energy (but not high-energy) photons. These attenuators are automatically inserted into, or removed from, a location in front of the detectors, depending on the detector counting rates at a given time. Imaging spectroscopy has been achieved over events spanning at least five orders of magnitude in flux, from GOES A-class events to the largest GOES X10 (and higher) events.

5. *Spectral Resolution.* Through the fine energy resolution ($\sim$1 keV FWHM below 100 keV to $\sim$5 keV at 2.2 MeV), *RHESSI* has the ability to accurately measure energy spectra as steep as ϵ^{-10} and to measure the widths of all detectable gamma-ray lines (except for the intrinsically very narrow neutron-capture gamma-ray line at 2.223 MeV).
6. *Detector Cross-Check.* The nine germanium detectors provide nine completely independent measures of the incident spectrum and so provide detailed cross checks and estimates of any systematic uncertainties in excess of those resulting from the counting statistics alone.
7. *Robustness of Data.* Since the basic *RHESSI* data consist of all time- and energy-tagged photon counts, the user can customize the energy range, angular resolution, and field of view of the reconstructed images as the scientific objectives dictate. In addition, existing data sets can be used without modification as increasingly sophisticated data analysis techniques are developed.
8. *Wide Range of Scientific Studies.* The availability of many image reconstruction techniques, many of which are optimized for the RMC imaging methodology, allows application to many scientific objectives.

Limitations

1. *Dynamic Range.* Since the *RHESSI* detectors necessarily measure the grid-throughput flux from *all* sources in the field of view, it is impossible to focus attention solely on a particular subregion of interest (or to exclude certain subregions). Further, the limited number of Fourier components sampled, coupled with systematic instrumental uncertainties, limit the ability to completely remove the side lobes from the reconstructed images. Thus, even with high signal-to-noise data, the observed signal at any given time contains overlapping information from multiple sources. This confusion of different sub-regions in the image limits *RHESSI*'s ability to detect weak sources in the presence of strong sources.

   The dynamic range of resolvable intensities is defined as the ratio of the surface brightness of the strongest source in an image to the weakest source that can be detected in that image. For *RHESSI*, it is limited for weak sources by the Poisson statistics of the number of recorded counts in the selected time and energy bins. For the strongest sources, the dynamic range is limited by imperfections in the grids and the systematic uncertainties in their measured characteristics and their relative alignments. It is also limited by uncertainties in the knowledge of the relative sensitivities of the nine germanium detectors. The design goal was that in the most favorable cases with sufficiently high counting rates, a dynamic range of 100:1 could be achieved. This is in comparison to $10^4 : 1$ for the AIA telescopes on SDO [124] and $\sim 10^6 : 1$ for the human eye. In practice, dynamic ranges of 10:1 are readily achieved by all image reconstruction algorithms and ~50:1 has been achieved in larger events with the best counting statistics. Significant improvements over this basic limitation are possible for extended sources using forward-fitting techniques in both image and spatial frequency spaces (Sects. 5.3 and 6.3).

2. *Pulse Pile-up.* Pulse pile-up results when two or more photons deposit their energy in a detector at nearly identical times and are erroneously recorded as a single photon with an energy equal to the sum of the energies of the two photons. Fortunately, use of the *RHESSI* movable attenuators means that the pile-up contribution is usually quite small [167]. However, since solar flare X-ray spectra fall off steeply with energy, even a small number of erroneous high-energy detections due to pulse pile-up can significantly affect the measured count-rate spectrum [46–48, 178].

   The contribution of pulse pile-up to the measured count-rate spectrum can be accommodated relatively straightforwardly in the *RHESSI* Object Spectral EXecutive (OSPEX) spectral analysis software[4] by using a forward method of computing the expected pileup contribution to each detector's spectrum. However, the effect on image morphology cannot be so easily taken into account, and pile-up can lead to ghost hard X-ray sources at high energies that are similar in shape and location to intense soft X-ray sources seen at lower energies. It can also lead to errors in estimating source dimensions, since such a determination depends on knowing the relative sensitivities of the different detectors at all times during the modulation cycle of each detector. For a given input photon spectrum, the average counting rates in the nine *RHESSI* detectors differ significantly because of their different absolute sensitivities and hence their effective relative sensitivities will vary because of pile-up as a function of phase of the modulation profiles. The difficulty in accurately correcting for the effects of pulse pile-up is magnified by two additional factors:

   - In an uncertain number of cases where pulse pile-up occurs, the events are vetoed by the on-board electronics and not included in the telemetry;
   - The detector output spectrum has a largely unknown noise component below the lower level discriminator (LLD) that is set at ~3 keV for most detectors most of the time. Thus, in those cases, the noise does not contribute appreciably to pulse pile-up. However, the LLD is set to ~10 keV for detector #7 and hence the unknown noise below this level can contribute significantly to the pile-up for that detector.

   Generally then, caution should be taken when making images at times during a flare when the total count rate in any one detector integrated over all energies is greater than $\sim 10^4$ counts s^{-1} at the peak of the modulation cycle for that detector.

[4] https://hesperia.gsfc.nasa.gov/rhessi3/software/spectroscopy/spectral-analysis-software/index.html.

Note that this only affects the front segments; the count rates in the rear segments are generally much lower than this so that pulse pile-up is not a problem.

3. *Detector Sensitivity.* The absolute sensitivity of the germanium detectors changes over long timescales (of order months). Hence absolute intensities could not be determined to better than ~10% early in the mission, and to an accuracy significantly worse than this as the detector performance degraded because of particle radiation damage later in the mission. This was partially compensated for by occasional detector annealing (heating and then re-cooling) that had to be carried out with increasing frequency as the mission progressed. As noted above, knowledge of the *relative*, rather than absolute, sensitivities of the different detectors is important for imaging. A method for correcting for the different relative detector sensitivities has been implemented: it uses the fact that all nine detectors see exactly the same incident photon flux, so that an approximate estimate of the photon flux can be calculated for each detector from the count rates averaged over an integral number of spacecraft rotations (using the diagonal elements of the response matrix for that detector). A correction factor is then calculated for each detector so that all photon fluxes estimated in this way are the same as the average of the original values for all detectors. This method allows the relative sensitivity of the nine detectors to be determined to within a few percent and significantly improves the quality of the images made with any of the reconstruction methods. It does not, however, correct for the different levels of pulse pile-up in the detectors due to their different absolute sensitivities.
4. *Variable Spacecraft Rotation Rate and Precession Angle.* Artifacts appear in the uncorrected counting rate light curves as a result of the variable spacecraft rotation rate and precession angle. These variations occur as the result of thermal day/night expansions and contractions, and of movements of the attenuators into and out of the detector lines of sight to the Sun. Fortunately, they are largely accounted for with the information from the aspect system and hence do not significantly affect the imaging capability.

3.4 *RHESSI* Imaging Example

An example illustrating *RHESSI*'s capability to image a simple asymmetric source with different extents in perpendicular directions was presented in [50] and is shown in Fig. 3.5. The 12–25 keV image shown in the top left panel of Fig. 3.5 was made during the GOES C7.9 class flare that occurred on 2011 September 25 in the National Oceanic and Atmospheric Administration (NOAA) active region 11302 at N12E50 [disk coordinates $X \simeq 700''$, $Y \simeq 151''$]), peaking at 03:32 UT. The white contours show the image made with the VIS_FWDFIT method (see Sect. 6.3) under the assumption that the source is a single elliptical Gaussian; the best fit to the visibilities $V(u, v)$ was obtained with FWHM dimensions of $(15.8 \pm 0.4)''$ in length and $(2.0 \pm 0.2)''$ in width. The counting rates and visibility amplitudes for each of the nine germanium detectors, shown in Fig. 3.5, illustrate how changes in the

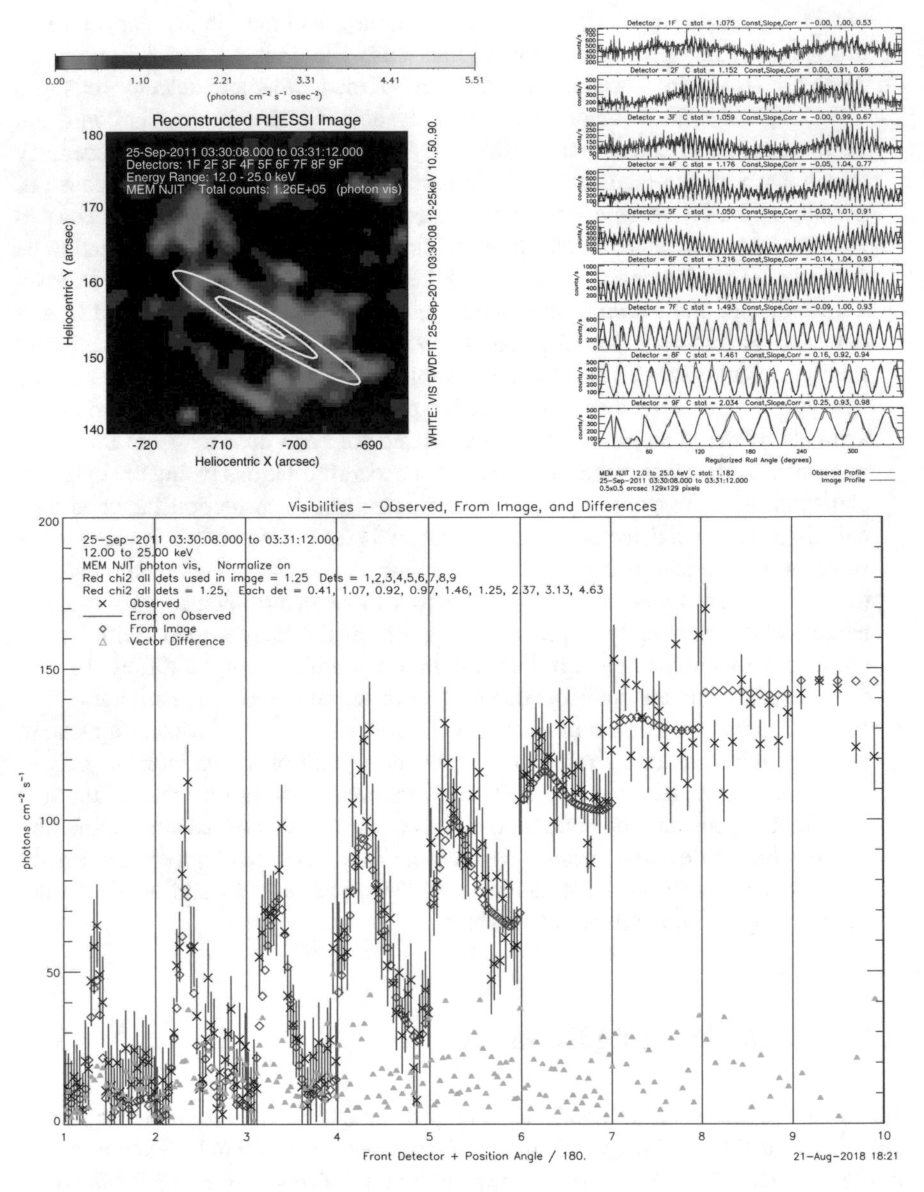

Fig. 3.5 *Top left*: *RHESSI* color image in the 12–25 keV energy range made using the MEM_NJIT (Sect. 6.5) reconstruction method. White contours show the extent of a best-fit elliptical Gaussian source determined with the VIS_FWDFIT (Sect. 6.3) procedure. *Top right*: Count rates (in red) in the front segment of each of the nine germanium detectors for the same time interval and energy band as for the image. These are plotted versus "regularized roll angle," defined as the spacecraft roll angle corrected for the offset between the spin axis and the mean subcollimator optical axis. Overlaid in black are the expected count rates predicted from the image shown to the left. *Bottom*: Visibility amplitudes plotted against position angle for each of the nine *RHESSI* detectors, shown as black crosses with $\pm 1\sigma$ error bars in blue. The amplitudes determined from the image reconstructed using MEM_NJIT (Sect. 6.5) are shown as red diamonds, with the vector difference amplitudes shown as green triangles. After [50], © AAS. Reproduced with permission.

modulations are produced by this relatively simple source and how inspection and analysis of these modulations leads to the reconstructed image. The top right panel in Fig. 3.5 shows the count rates for each detector separately plotted as a function of "regularized roll angle," defined as the spacecraft roll angle corrected for the offset between the spin axis and the mean subcollimator optical axis. To build up statistics, the count rates are "stacked," i.e., summed modulo the spacecraft spin period ($\sim$4 s) for the 64 s interval duration used for the image.

Significant sinusoidal modulation is evident in the plots for all detectors. The detectors behind the subcollimators with the coarser grids (#5 to #9) show modulation at all roll angles, while the detectors behind the finer grids (#1 to #4) show modulation over two limited ranges in regularized roll angle ($\sim$90–150° and $\sim$280–300°). This is the expected modulation signal for an asymmetrical source with a width much smaller than its length, oriented as shown in the top left panel. The agreement between the measured and predicted rates is an indication of how well the reconstructed image matches the data. This agreement is quantified with the Cash [34] or C-statistic, given as a separate value for each detector and as an overall value for all detectors together.

Another way of displaying the modulation in the different detectors is shown in the bottom panel of Fig. 3.5, sometimes called a "Hurford" plot after its originator. Here, the amplitudes of the visibilities for each detector are plotted as a function of position angle, defined as the spatial direction of each grid response referenced to solar North. This is essentially the same as plotting the amplitudes of the count-rate oscillations seen in the top right panel, except that the data are plotted for only a half rotation from 0° to 180° (with the second half rotation assumed to be identical and hence added to the first, an option known as "combine conjugates"). Here again, the characteristics of an asymmetric source are dramatically evident, with the peaks in the visibility amplitudes for each detector showing the position angle of the smallest dimension of the source and the valleys showing the position angle of the largest dimension. Since the peak is evident even in detector #1, we know that the smallest source dimension must be close to the 2″ resolution of the finest grids. The count rates in the valleys are similar in detectors #1, 2, and 3, with detector #5 having $\sim$50% of the modulation amplitude of detectors #8 and 9. This indicates that the longest source dimension must be similar to the 20″ resolution of detector #5, i.e., similar to the value found from VIS_FWDFIT. From this analysis, another quantitative indication of how well the image fits the count rates is given by the reduced χ^2 values computed from the measured and predicted visibility vectors weighted by the statistical uncertainties. These values are listed in the figure for each detector and for all detectors together.

3.5 SSW and the *RHESSI* GUIs

Although the ability to reconstruct X-ray images from the *RHESSI* observations for individual solar flares is computationally challenging, a large software development effort by the *RHESSI* team has resulted in extensive user-friendly capabilities for fully exploiting the imaging spectroscopy capabilities of the instrument. As pointed out in [168], the main challenges that had to be overcome were the large volume of data from the instrument (which records the time, detector number, and energy of each detected count, amounting to hundreds of millions of events for some flares), the general unfamiliarity of the user community with Fourier-transform imaging, and the need to compare *RHESSI* observations with scientifically important context information from instruments operating in other wavelength domains. With the advances in computing power and memory capacity in the almost 20 years since the original software was written, the first challenge is now much less acute. The computational task is to take this basic ("Level-0") data, together with calibration and pointing ("aspect") information, to produce the modulated count rates from each of the two segments of all nine germanium detectors. These calibrated counting rates are then used to produce light curves, spatially integrated spectra, and images.

The generation of the vast array of computational routines now available for the scientific analysis of *RHESSI* observations was greatly facilitated by the open data and software policy established by the Principal Investigator team. All data from the spacecraft were made publicly available on line, generally within 24 h of its receipt at the Mission Operations Center at the Space Sciences Laboratory of the University of California at Berkeley. The software, written in the Interactive Data Language (IDL, Harris Geospatial Solutions, Boulder, Colorado), is incorporated into the SolarSoftWare (SSW) system [74] and makes full use of all the existing utilities and libraries therein.[5]

All of *RHESSI*'s IDL data analysis procedures are based on object-oriented design concepts, in which the software is divided into independent parts, called *objects*, that interact with standard interfaces. While all the procedures can be run from the IDL command line, several extensive, more user-friendly capabilities have been developed using two Graphical User Interfaces (GUIs), both accessible from the IDL command line after the SSW library has been installed. The first GUI is initiated by typing[6] "hessi" at the IDL command line. It allows the user to show pre-made quick-look lightcurves, generate customized light curves and spectra, and make *RHESSI* images using any of the available reconstruction routines. An additional invaluable capability provides access to synoptic data from other instruments, allowing *RHESSI* images to be co-aligned and contours overlaid on co-temporal images taken at other wavelengths including optical, radio, EUV, and soft X-ray.

[5] Detailed documentation, including instructions for installing SSW, is available online at https://hesperia.gsfc.nasa.gov/rhessi/software/installation/index.html. Information on how to access and use all *RHESSI* data and software is available at https://hesperia.gsfc.nasa.gov/rhessi/.

[6] The original name for *RHESSI* was HESSI. The leading "R" was added shortly after launch in tribute to Reuven Ramaty, one of the pioneers of gamma-ray astronomy and a major contributor to establishing the scientific justification for *RHESSI*'s high resolution gamma-ray spectroscopy. Sadly, he died just months before *RHESSI* was launched.

The second, or OSPEX GUI is initiated by typing "obj=ospex()" at the IDL command line. It provides access to the extensive spectroscopy and imaging spectroscopy tools available for analysis of *RHESSI* data, and also X-ray and gamma-ray spectral data from other instruments. This GUI allows the user to read data files in the standard Flexible Image Transport System (FITS) format that contain either spectral information alone or image cubes made up of images in multiple energy and/or time bins. These FITS files can be either those made with the GUI for *RHESSI* data analysis (also called the HESSI GUI) or by software specific for other instruments. After subtracting any non-solar background, the spectral information is converted from the measured count rates to photon fluxes using the full instrument response matrix including the off-diagonal elements. Since this conversion depends on the unknown form of the solar flare spectrum, a forward fitting method is used. A specific spectral function is assumed and an iterative procedure is followed to obtain the values of the function parameters that give the best agreement with the measured count rates when the function is folded trough the instrument response matrix. Many functional forms are available including those predicted for X-ray bremsstrahlung emission from thermal plasma and from flare accelerated electrons with power-law spectra. This spectral analysis can be carried out on either the spatially integrated flux without the need to make any images or on specific selected regions of the images. Generally, even with the high counting rates during big flares, the imaging dynamic range limitations mean that spectra of only two or three isolated regions can be analyzed separately.

In addition to the spectral fitting of different spectral forms, OSPEX also allows an albedo component to be added (see, e.g., [15, 102, 106]). This corresponds to X-rays directed downwards from the primary source towards the solar surface, but then reflected back to the observer. When spectra from individual detectors are analyzed, a pulse pileup component can also be added and the default detector resolution and energy calibration can be modified to better fit the data.

The availability of these *RHESSI* GUIs has greatly expanded the number of people who have been able to carry out scientific analysis of *RHESSI* observations, including the almost routine generation of calibrated light curves, spectra, and images for any given flare. Even imaging spectroscopy is possible with the OSPEX GUI.

3.6 *STIX* Design/Brief History of Concept Development

The *STIX* instrument on the European *Solar Orbiter* mission, launched on 2020 February 9, is shown in Fig. 3.6 and described in detail in [120]; the description in the present section is derived largely from that paper. *STIX* uses a Fourier-transform imaging technique with multiple bigrid modulation collimators, similar to that used by HXT [113] on the Japanese Yohkoh mission, and the Hard X-ray Imager to be flown on the upcoming (at the time of writing) Chinese ASOS mission [75, 208]. The basic technique is related to that used for *RHESSI* but, since

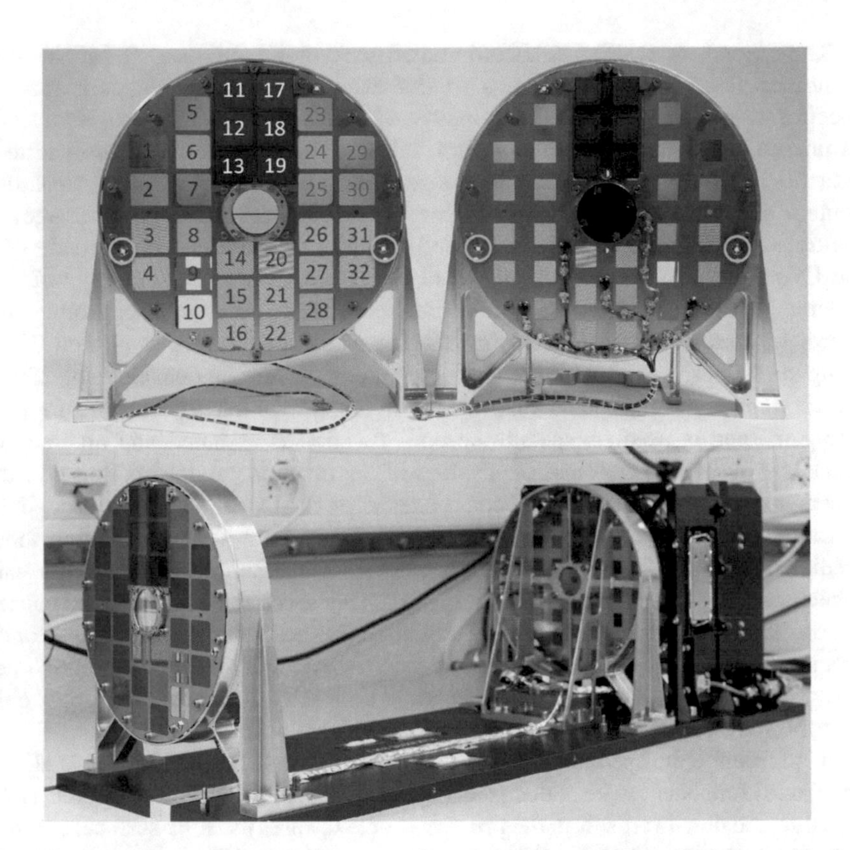

Fig. 3.6 *Top*: Photograph of *STIX* flight-spare front (left) and rear (right) grids, mounted in their support frames. The colored numbers relate to the grid pitch, with the colors indicating grids that were etched on the same tungsten foils. Individual front grids have usable areas of 22 × 20 mm (front) while the rear grids are slightly smaller at 13 × 13 mm. *Bottom*: Photograph of the *STIX* instrument layout showing the imager with two grid assemblies separated by 55 cm. The Cadmium Telluride (CdTe) detectors, together with a movable X-ray attenuator and the instrument electronics, are in the black box behind the rear grids. After [120], with permission from the EDP Sciences for European Southern Observatory.

the Solar Orbiter is a 3-axis stabilized spacecraft with limited mass, power, and telemetry resources, a modified approach was adopted. This was made possible by the development of an advanced etching technique that allowed fine grids to be made that were similar to those used for *RHESSI* but with multiple pitches on a single tungsten sheet. The *RHESSI* finer grids (#1, 2, 3, and 4), each 9 cm in diameter, were made up of thin tungsten sheets (except for the finest grid made of molybdenum) stacked on top of one another to form the complete grid. Each sheet had fine parallel slits photo-chemically etched through to make a grid with a given pitch defined as the distance between adjacent slits. For *STIX*, Mikro Systems in Charlotte, VA (http://www.mikrosystems.com/) built up multiple grids, each 1 cm square and with a different pitch, to be etched into single tungsten foils. Stacking multiple foils produced the flight grids of the required thickness to absorb X-rays up to the 150 keV end of the *STIX* energy range.

The *STIX* instrument consists of two X-ray opaque grid assemblies separated by 55 cm. Each of these grid assemblies is divided into 32 grids with a corresponding set of 32 coarsely-pixelized detectors located behind the rear assembly. A pair of front and rear grids and its corresponding detector is termed a "subcollimator." Among these 32 subcollimators, one is used as a coarse flare locator and another is used for background and unsaturated large-flare measurements. Within each one of the remaining 30 subcollimators, the two grids contain a large number of parallel, equally-spaced slits but the slits in the front and rear grids differ slightly in pitch and/or orientation. These differences are selected so that, for each subcollimator, the combined X-ray transmission of the grid pair forms a large-scale Moiré pattern on the detector with a period equal to the detector width and oriented to be parallel to a detector edge. The amplitude and phase of this pattern are measured using four independent parallel strips in the detector. They are very sensitive to the angular distribution of the incident X-ray flux and its orientation with respect to the instrument optical axis. Thus, the high-angular resolution X-ray imaging information is encoded into a set of large-scale spatial distributions of counts in the detectors. The grid design provides the imaging information in the form of a set of up to 30 angular Fourier components (i.e, visibilities) of the source distribution. In contrast to *RHESSI*, where over 100 Fourier components were measured every 4 s as the spacecraft rotated at 15 rpm, the *STIX* Fourier components are all available on as short a time scale as the count rates allow—less than 1 s for the larger flares. The sets of measured visibilities are used on the ground to reconstruct images of the X-ray source using software already developed for *RHESSI* with minimal modifications.

3.7 The *STIX* Imaging Concept

A detailed formal description of the *STIX* imaging concept is given in [77, 130]. This section uses the content of those papers in order to provide an overview of the image formation process in *STIX*.

In order to describe the signal formation process in *STIX*, we introduce four parallel two-dimensional coordinate systems $(\mathbf{x}, \mathbf{z}, \mathbf{q}, \mathbf{s})$ determined, respectively, by the image of the Sun in the plane of the sky, the front grid, the rear grid, and the detector surface. From a point $\mathbf{x} = (x, y)$ on the solar disk, a photon flux $\phi(\mathbf{x})$ passes through the front grid (at distance S from the Sun) at the point $\mathbf{z} = (z_1, z_2)$, then through the rear grid (at distance L_1 from the front grid) at the point $\mathbf{q} = (q_1, q_2)$, and finally impacts the detector surface (a distance L_2 from the rear grid) at the point $\mathbf{s} = (s_1, s_2)$ (see Fig. 3.7).

According to the current release of the *STIX* simulation software, the coordinate system on the Sun, represented by $\mathbf{x}$, is flipped with respect to the coordinate systems on the grids and on the detector. Taking this into account, standard relations between similar triangles in Fig. 3.7 lead to

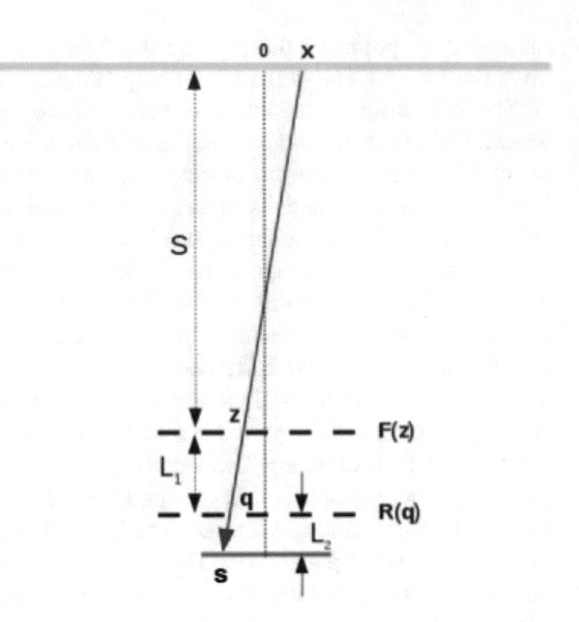

Fig. 3.7 Schematic representation of the signal transmission process through one of the *STIX* sub-collimators. The red line is the path traveled by a photon emitted from the point **x** on the Sun (yellow surface) and reaching the point **s** on the detector (in green). Each coordinate plane has two directions (e.g., (q_1, q_2) for the rear grid), corresponding respectively to displacements normal to, and in the plane of, the figure.

$$\mathbf{z} = \mathbf{s} - \frac{(\mathbf{x} + \mathbf{s})\,(L_1 + L_2)}{S + L_1 + L_2} \simeq \mathbf{s} - \mathbf{x}\,\frac{L_1 + L_2}{S} \tag{3.6}$$

and

$$\mathbf{q} = \mathbf{s} - \frac{L_2\,(\mathbf{x} + \mathbf{s})}{S + L_1 + L_2} \simeq \mathbf{s} - \mathbf{x}\,\frac{L_2}{S}\,, \tag{3.7}$$

which gives $\mathbf{z} = \mathbf{z}(\mathbf{x}, \mathbf{s})$ and $\mathbf{q} = \mathbf{q}(\mathbf{x}, \mathbf{s})$. The approximations in these two equations neglect terms of the order of centimeters with respect to terms of the order of the distance to the Sun, or $\sim 10^8$ km, and so are extremely accurate, to about 1 part in 10^{13}.

The transmission functions $F(\mathbf{z})$ and $R(\mathbf{q})$ through the front and rear grids, at points **z** and **q** respectively, can be modeled as square-wave functions and therefore represented in terms of their Fourier series:

$$F(\mathbf{z}) = \frac{1}{2} + \frac{2}{\pi} \sum_{m=1,3,5,\cdots} \frac{1}{m} \sin\left(2\pi m \cdot (\mathbf{z} + \mathbf{t})\right) \tag{3.8}$$

and

$$R(\mathbf{q}) = \frac{1}{2} + \frac{2}{\pi} \sum_{j=1,3,5,\cdots} \frac{1}{j} \sin\left(2\pi j \cdot (\mathbf{q} + \mathbf{t})\right)\,, \tag{3.9}$$

where the constant vector $\mathbf{t} = (0, t)$ accounts for a possible translation in the y direction (i.e., in the plane of Fig. 3.7) of the grid pair with respect to the origin. Using Eqs. (3.6) and (3.7), Eqs. (3.8) and (3.9) can be represented in terms of $\mathbf{x}$ and $\mathbf{s}$ as

$$F(\mathbf{x}, \mathbf{s}) = \frac{1}{2} + \frac{2}{\pi} \sum_{m=1,3,5,\cdots} \frac{1}{m} \sin\left(2\pi m \cdot \left(\mathbf{s} - \mathbf{x}\frac{L_1 + L_2}{S} + \mathbf{t}\right)\right) \tag{3.10}$$

and

$$R(\mathbf{x}, \mathbf{s}) = \frac{1}{2} + \frac{2}{\pi} \sum_{j=1,3,5,\cdots} \frac{1}{j} \sin\left(2\pi j \cdot \left(\mathbf{s} - \mathbf{x}\frac{L_2}{S} + \mathbf{t}\right)\right) , \tag{3.11}$$

and the global transmission function for each *STIX* sub-collimator is given by

$$T(\mathbf{x}, \mathbf{s}) = F(\mathbf{x}, \mathbf{s})\, R(\mathbf{x}, \mathbf{s}) . \tag{3.12}$$

Knowledge of the global transfer function $T(\mathbf{x}, \mathbf{s})$ allows the computation of the number of counts recorded by each pixel[7] in a given *STIX* detector. The probability $\tau_n(\mathbf{x})$ that a photon starting from position $\mathbf{x}$ on the Sun reaches pixel n in a specific detector is given by

$$\tau_n(\mathbf{x}) := \int_{P_n} T(\mathbf{x}, \mathbf{s})\, d\mathbf{s} \quad n = -2, -1, 0, 1 \tag{3.13}$$

and so the overall number of counts in that pixel is given by

$$\int_S I(\mathbf{x})\, \tau_n(\mathbf{x})\, d\mathbf{x} \quad n = -2, -1, 0, 1 , \tag{3.14}$$

where $I(\mathbf{x})$ is the photon flux entering the telescope. For a given source coordinate $\mathbf{x}$, Eq. (3.14) defines the Moiré fringe pattern on each detector.

It can be shown [77] that an approximate estimate for the number of counts recorded by the nth pixel in each *STIX* detector is

$$C_n \simeq M_0\, V(\mathbf{0}) + M_1\, e^{i\pi n/2}\, e^{i\pi/4}\, V(-\boldsymbol{\xi}) + M_1\, e^{-i\pi n/2}\, e^{-i\pi/4}\, V(\boldsymbol{\xi}) , \tag{3.15}$$

where

$$V(\boldsymbol{\xi}) = \int I(\mathbf{x})\, e^{i2\pi\boldsymbol{\xi}\cdot\mathbf{x}}\, d\mathbf{x} \tag{3.16}$$

[7] The four pixels in each detector are numbered as -2, -1, 0 and 1, so that the quantity $e^{i\pi n/2}$, used in the sequel, takes on the purely real or imaginary values -1, $-i$, 1, and i.

and each vector

$$\boldsymbol{\xi} = \mathbf{k}^f \, \frac{L_1 + L_2}{S} - \mathbf{k}^r \, \frac{L_2}{S} \tag{3.17}$$

in the spatial frequency space is associated with a specific sub-collimator. In fact, for each sub-collimator the front (f) and rear (r) grids are characterized by pitches p^f, p^r and orientation angles α^f, α^r, in such a way that

$$\mathbf{k}^f = \left(\frac{\cos \alpha^f}{p^f}, \frac{\sin \alpha^f}{p^f} \right); \qquad \mathbf{k}^r = \left(\frac{\cos \alpha^r}{p^r}, \frac{\sin \alpha^r}{p^r} \right). \tag{3.18}$$

Since each sub-collimator has a unique grid orientation and a unique pitch, each $\boldsymbol{\xi}$ corresponds to a unique point in the spatial frequency plane.

As discussed above, each *STIX* detector contains four pixels corresponding to $n = -2, -1, 0, 1$. The number of counts detected by each of the four pixels are

$$A \equiv C_{-2} \simeq M_0 \, V(\mathbf{0}) - M_1 \, e^{i\pi/4} \, V(-\boldsymbol{\xi}) - M_1 \, e^{-i\pi/4} \, V(\boldsymbol{\xi}) \, , \tag{3.19}$$

$$B \equiv C_{-1} \simeq M_0 \, V(\mathbf{0}) - i \, M_1 \, e^{i\pi/4} \, V(-\boldsymbol{\xi}) + i \, M_1 \, e^{-i\pi/4} \, V(\boldsymbol{\xi}) \, , \tag{3.20}$$

$$C \equiv C_0 \simeq M_0 \, V(\mathbf{0}) + M_1 \, e^{i\pi/4} \, V(-\boldsymbol{\xi}) + M_1 \, e^{-i\pi/4} \, V(\boldsymbol{\xi}) \, , \tag{3.21}$$

and

$$D \equiv C_1 \simeq M_0 \, V(\mathbf{0}) + i \, M_1 \, e^{i\pi/4} \, V(-\boldsymbol{\xi}) - i \, M_1 \, e^{-i\pi/4} \, V(\boldsymbol{\xi}) \, . \tag{3.22}$$

Combining these, we find that

$$C - A \simeq 4M_1 \mathrm{Re} \left(e^{-i\pi/4} \, V(\boldsymbol{\xi}) \right) \tag{3.23}$$

and

$$D - B \simeq 4M_1 \mathrm{Im} \left(e^{-i\pi/4} \, V(\boldsymbol{\xi}) \right) \, , \tag{3.24}$$

where we have exploited the fact that $V(-\boldsymbol{\xi}) = \overline{V(\boldsymbol{\xi})}$, i.e., the complex conjugate of $V(\boldsymbol{\xi})$. Equations (3.23) and (3.24) give the (complex) visibility at the spatial frequency point $\boldsymbol{\xi} = (u, v)$ in terms of the pixel counts (A, B, C, D):

$$V(\boldsymbol{\xi}) \simeq \frac{1}{4M_1} \left[(C - A) + i \, (D - B) \right] e^{i\pi/4} \, . \tag{3.25}$$

This last equation shows how the four pixel counts C_n; $n = -2, -1, 0, 1$ in each detector can be used to yield the visibility at the frequency domain points $\boldsymbol{\xi}$ given, for each detector geometry, by Eqs. (3.17) and (3.18). Since $V(-\boldsymbol{\xi}) = \overline{V(\boldsymbol{\xi})}$, there

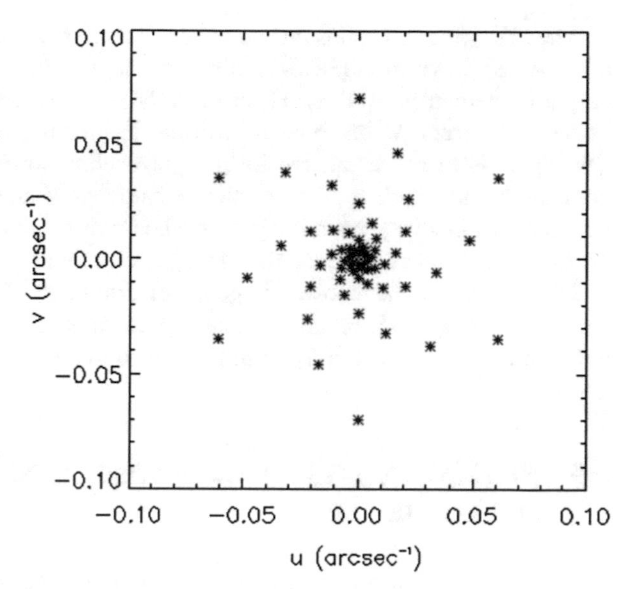

Fig. 3.8 Representation of the sampling of the spatial frequency plane at $S = 1$ Astronomical Unit (AU) in *STIX*, where we have used the fact that the property $V(-\xi) = \overline{V(\xi)}$ allows the duplication of the 30 spatial frequency samples.

are 60 sampled points overall. The *STIX* sub-collimators are engineered with 10 different pitches, increasing in a geometric series with ratio 1.43 from 38 μm to 953 μm, and nine orientation angles uniformly increasing in 20° steps from 10° to 170°. As a result, the sampled points in the (u, v) plane are arranged in a pattern of six spiral arcs, shown in Fig. 3.8. The errors due to the truncation of the Fourier series in Eq. (3.15) are in the range 2–7.5% for all sub-collimators [77] with the errors being larger for sub-collimators with smaller pitches.

3.8 *STIX* Software

The *STIX* imaging software is organized in two packages, one concerning the simulation of *STIX* counts and visibilities and the other one containing the data analysis tools.

The simulation package utilizes Monte Carlo techniques in order to produce count signals at different energies on the basis of the signal formation model described in the previous section. Given an assumed shape for the X-ray source at the Sun, the Monte Carlo scheme iteratively and randomly extracts a point within the field of view by means of a probability distribution in accordance with the source shape. It then randomly selects a direction for the emitted X-rays from this point and follows their paths through each grid pair and records the corresponding hits in the appropriate pixel of each detector. Finally, visibilities associated with each detector are computed according to Eq. (3.25).

The *STIX* mission relies on a legacy of image reconstruction software that has been passed down from previous missions and, in particular, by the computational corpus implemented for the analysis of *RHESSI* counts and visibilities that is hosted in the SSW tree. At the time of writing, the main effort of the *STIX* software developers is to render all the imaging algorithms available to the solar physics community independent of the particular hardware characteristics of the observing telescope. In this way, each imaging method will be able to process any incoming set of visibilities and produce the corresponding image cube $I(x, y; \epsilon)$, no matter which instrument was involved in generating those visibilities. Chapters 4 and 6 of the present book describe the rationale behind this battery of algorithms, although new approaches will certainly be implemented in the near future.

3.9 *RHESSI* vs. *STIX*: A Comparison of Strengths and Limitations

The scientific rationales for *RHESSI* and *STIX* are both nominally characterized by the same objective: the understanding of how energy release, and in particular electron acceleration, occurs during solar flares. Indeed, the performances of the two instruments do present some similarities, as summarized in Table 3.1. For instance, for both instruments, the low energy limit corresponds to a thermal population with a temperature above ~10 MK, so that diagnostics of the hottest plasmas is feasible for both instruments. However, *RHESSI* observations were characterized by a spectral resolution higher than that associated with *STIX* and included detections of gamma-rays. Because of its smaller detectors and different electronics, *STIX* does not suffer as much from detector-saturating pulses from cosmic ray interactions, pulses that resulted in an ~20% loss in data from all but one of the *RHESSI* front segments.

RHESSI had significantly better coverage of the (u, v) plane (see Fig. 3.9). However, the *RHESSI* spacecraft rotation period of ~4 s and the data stacking process typically used limited its temporal resolution to a value that is coarser than that provided by *STIX*. In this context it should be noted that the trajectory of the Solar Orbiter mission on which the *STIX* instrument is located involves passes much closer to the Sun than the nearly-constant 1 AU provided by *RHESSI*'s Earth

Table 3.1 Comparison of specifications of the *RHESSI* and *STIX* instruments

	RHESSI	*STIX*
Energy range	3 keV–17 MeV	4–150 keV
Spectral resolution	1–2 keV (below 1 MeV)	1–15 keV (depending on energy)
Time resolution	>4 s	>1 s
Spatial resolution	2–180 $''$ (depending on energy)	>7 $''$ (depending on energy)

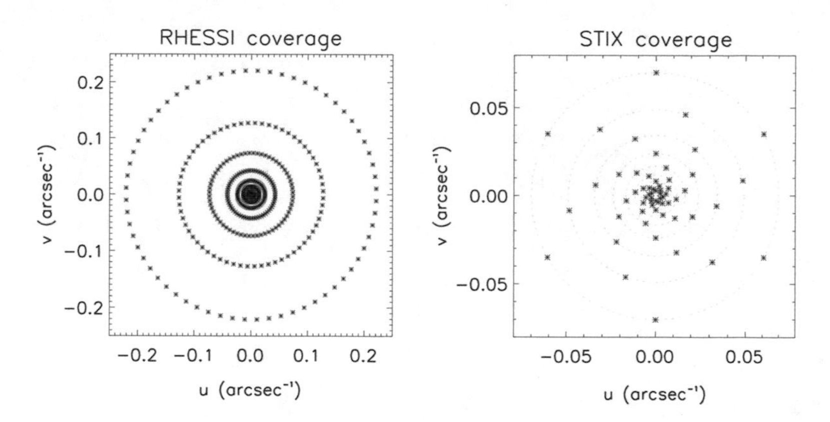

Fig. 3.9 Representation of the sampling of the spatial frequency or (u, v) plane for *RHESSI* (left panel) and *STIX* (right panel). After [120], used with permission of the EDP Sciences for European Southern Observatory.

orbit trajectory; indeed, in a series of encounters spanning from 2023 to 2028, Solar Orbiter will pass as close as 0.28 AU to the Sun. At such distances, a given angular resolution represents a much smaller spatial scale on the Sun: $1''$ corresponds to a distance of only $\sim$200 km (versus $\sim$725 km for *RHESSI*). Thus, although the finest angular resolution achievable by *STIX* is approximately three times poorer than *RHESSI* (a maximum spatial frequency of $\sim$0.07 arcsec^{-1} versus $\sim$0.22 arcsec^{-1} for *RHESSI*—Fig. 3.9), this factor is almost exactly compensated by the closer distance to the Sun. For example, at a spacecraft distance of 0.28 AU from the Sun, an angular resolution of $\alpha = 1/2R = 1/(2 \times 0.07) \simeq 7.1''$ corresponds to a spatial scale of $\sim 7.1'' \times 200$ km/arcsec $\simeq$1500 km, comparable to the finest scale of 1700 km ($2.3'' \times 725$ km/arcsec) for *RHESSI*. Thus, although the finest *angular* resolution of *STIX* is significantly coarser than that of *RHESSI*, at close approaches to the Sun its *spatial* resolution capability is very similar.

Notwithstanding the above technical differences, the main *operational* difference between the two instruments lies in the fact that *RHESSI* was the only instrument on its spacecraft, whereas *STIX* is integrated into the Solar Orbiter cluster with a total of ten instruments and 21 sensors. This allows observational campaigns in which *STIX* data are combined with both remote sensing and *in situ* observations from the other instruments. From this perspective, it is probably possible to state that *RHESSI* is genuinely a solar physics telescope, while *STIX* brands itself more as a heliophysics instrument.

Chapter 4
Image Reconstruction Methods

Abstract In this chapter we discuss the mathematical essence of the problem of reconstructing a two-dimensional image from data that does not conform to a straightforward "pixel-by-pixel" approach to the imaging process. Early on, we point out that the reconstruction of images from such data is plagued by several problems: (1) a formal solution may not exist at all, (2) solutions that are obtained may not be unique, and (3) the reconstructed images are subject to considerable uncertainty in the presence of data noise and uncertainties in the instrument parameters. In the next chapters, various methods to mitigate these issues in a physical context are described; each method is typically characterized by a parameter or set of parameters that is selected by the user in order to produce desirable properties of the reconstructed image or to optimize a reconstructed image according to some defined norm.

4.1 The Essence of the Image Reconstruction Problem

Consider a point source located at position (x, y) on the solar disk, corresponding to polar coordinates (θ, ϕ) with respect to the satellite position. Essentially any imaging device reproduces that point source as a blurred feature $g(x_i, y_i)$ on the image plane. Since any source can be represented as a superposition $I(x, y)$ of point sources, and assuming that the imaging system is linear, the corresponding image $g(x_i, y_i)$ is mathematically represented by

$$g(x_i, y_i) = \int_{-\infty}^{\infty} \int_{-\infty}^{\infty} I(x, y)\, K(x, x_i; y, y_i)\, dx\, dy\,, \tag{4.1}$$

where $K(x, x_i; y, y_i)$ is the kernel function, which specifies the image pattern $g(x_i, y_i)$ produced by a point source of unit intensity located at (x, y). For most instruments, $K(x, x_i; y, y_i)$ is a function that is independent of time; however, in the case of *RHESSI* the spinning of the instrument around the spacecraft rotation axis implies that a point source produces a characteristic temporal modulation pattern

M. Piana et al., *Hard X-Ray Imaging of Solar Flares*,
https://doi.org/10.1007/978-3-030-87277-9_4

$K(x, x_i; y, y_i, t)$. Thus, the corresponding image must be modeled as a function of both position and time:

$$g(x_i, y_i, t) = \int_{-\infty}^{\infty} \int_{-\infty}^{\infty} I(x, y)\, K(x, x_i; y, y_i; t)\, dx\, dy\, . \tag{4.2}$$

For both models (4.1) and (4.2) the essence of the image reconstruction problem is to determine the source intensity $I(x, y)$ from noisy observations of $g(x_i, y_i)$ or $g(x_i, y_i, t)$ and knowledge of the kernel functions $K(x, x_i; y, y_i)$ or $K(x, x_i; y, y_i; t)$.

Due to the smoothing effect of the kernel function, even a modest level of noise in observations leads to a considerably amplified level of uncertainty in the source function. Further, in general the kernel function is only partially determined; not accounting for this limited knowledge of K may lead to the appearance of side lobes on the restored image $I(x, y)$.

4.1.1 Count-Based Versus Visibility-Based Imaging

In models (4.1) and (4.2) the signal content in each pixel (x_i, y_i) is given by the number of counts produced in detectors by the incoming photon flux and is therefore characterized by Poisson noise. However, as explained in Chap. 3, certain classes of hard X-ray imaging devices are designed in such a way that the counts recorded in the detector pixels can be re-arranged in the form of a sparse and finite set of visibilities. The model equation for these instruments is

$$V(u_j, v_j) = \int_{-\infty}^{\infty} \int_{-\infty}^{\infty} I(x, y)\, e^{i2\pi(xu_j + yv_j)}\, dx\, dy\ ;\ j = 1, \ldots, N\, , \tag{4.3}$$

where $\{(u_j, v_j)\}_{j=1}^{N}$ is the set of N locations in the frequency domain at which visibilities are measured. Equation (4.3) can be written in the more compact form

$$V(u, v) = (\chi \mathcal{F} I)\,(u, v), \tag{4.4}$$

where $\chi = \chi(u, v)$ is the characteristic function that samples the spatial frequency (u, v) domain at the points where the visibilities are measured and $\mathcal{F}$ denotes the Fourier transform operator. By applying an inverse Fourier transform to both sides of Eq. (4.4) one obtains

$$g(x, y) = (K * I)\,(x, y)\, , \tag{4.5}$$

where

$$g(x, y) = \frac{1}{(2\pi)^2} \int_{-\infty}^{\infty} \int_{-\infty}^{\infty} V(u, v)\, e^{-2\pi i (xu+yv)}\, du\, dv \tag{4.6}$$

and

$$K(x, y) = \frac{1}{(2\pi)^2} \int_{-\infty}^{\infty} \int_{-\infty}^{\infty} \chi(u, v)\, e^{-2\pi i (xu+yv)}\, du\, dv\,. \tag{4.7}$$

Equations (4.3) through (4.7) show that even rearranging the count data into Fourier components (as is done with *RHESSI* and *STIX*) does not change the intrinsic characteristic of the mathematical model for image reconstruction: it remains that of a linear integral equation of the first kind [44]. Clearly, the role of the function $K(x, y)$ is crucial: it encodes the way the structural properties of the imaging device impact the signal formation process. Therefore, its shape, together with knowledge of the statistical properties of the noise affecting the measurements, provide limits on the amount of information on the X-ray source function that can be restored from the recorded data.

4.1.2 Point Spread Functions

A point spread function (PSF) is defined as the instrument response to a unit-intensity point source. Plots of the PSF for each of the nine *RHESSI* detectors, both separately and together, are shown in the left panel of Fig. 4.1. For comparison,

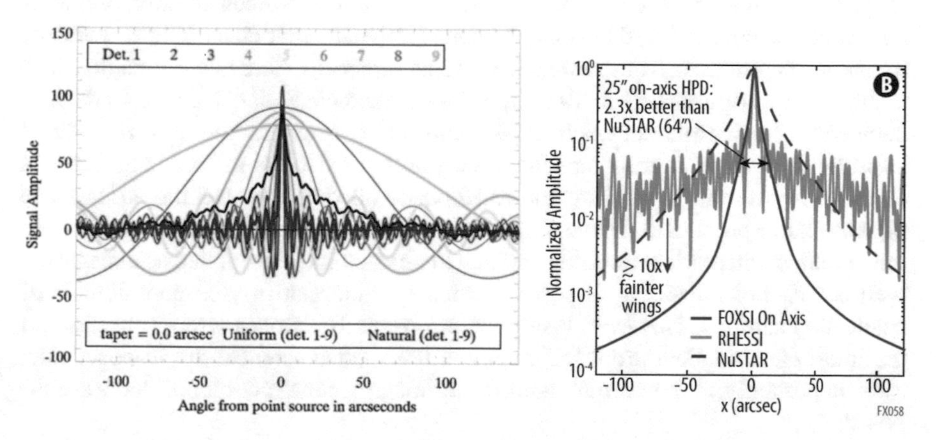

Fig. 4.1 Left: Plot of the point spread function for each of the nine *RHESSI* subcollimators. The relative signal amplitude in a given image pixel is plotted as a function of angular distance of the pixel from the location of a point source. Also shown is the PSF for all nine detectors used together with either uniform or natural weighting. Right: PSF of the proposed FOXSI telescope in comparison to the *RHESSI* and NuSTAR PSFs.

the PSFs of two focusing optics X-ray telescopes (the Focusing Optics X-ray Solar Imager (FOXSI) [119] and NuSTAR [85]) are shown in the right panel of Fig. 4.1. Note that *RHESSI* has significantly finer angular resolution, thanks to the finest subcollimators, but has higher and more complex side lobes than the focusing telescopes.

There are two ways of characterizing the PSF for any given instrument: the FWHM and the half-power diameter (HPD) specified as the offset angle within which half the power of the source is recorded. The FWHM is useful for specifying the finest possible angular resolution, while the HPD characterizes the amplitude of the wings of the distribution that determine the possible dynamic range that can be achieved.

Figure 4.2 illustrates a direct comparison between the PSFs associated to *RHESSI* and *STIX*. These plots show that the *STIX* PSF is characterized by significantly more oscillations than is the *RHESSI* PSF. Further, the FWHM of the *RHESSI* PSF is approximately a half that of the *STIX* PSF, which implies that *RHESSI* is able to achieve an angular resolution $\sim$2 times finer than *STIX*. Of course, when *STIX* is at 0.2 AU, it will have similar *spatial* resolution to that of *RHESSI* in Earth orbit at 1 AU.

4.2 The Ill-posedness of the Image Reconstruction Problem

As discussed in Chap. 3, instruments such as *RHESSI* and *STIX* record imaging information indirectly through techniques that essentially provide spatial Fourier transforms of the source structure. Reconstructing the image properties from such data is a challenging task, for several reasons. These include a crucial aspect that is related to a pathology associated with essentially all image reconstruction problems and which accounts for the general interest of mathematicians in the world of image reconstruction, from remote-sensing astrophysics to medical imaging. In a technical report published in 1923, the French mathematician Jacques Hadamard decided to classify all mathematical problems into two general families. He defined as *well-posed* all equations for which the solution exists, is unique, and depends continuously on the input data, and as *ill-posed* all problems lacking at least one of these three properties. Interestingly enough, the influential Hadamard believed that all the mathematical questions related to physics and the applied sciences were well-posed, thus relegating ill-posed problems to the relatively obscure domain of academic exercises. However, history soon proved Hadamard wrong: it turns out that most mathematical problems arising in the applied sciences are ill-posed and, most importantly in the current context, *all* image reconstruction problems are ill-posed.

The three pathologies of ill-posedness, i.e., nonexistence, non-uniqueness, and instability of the solution, all impact the image reconstruction problem in different ways. Indeed (and almost paradoxically), the fact that the solution of the image reconstruction problem does not even exist for every set of input data turns out *not*

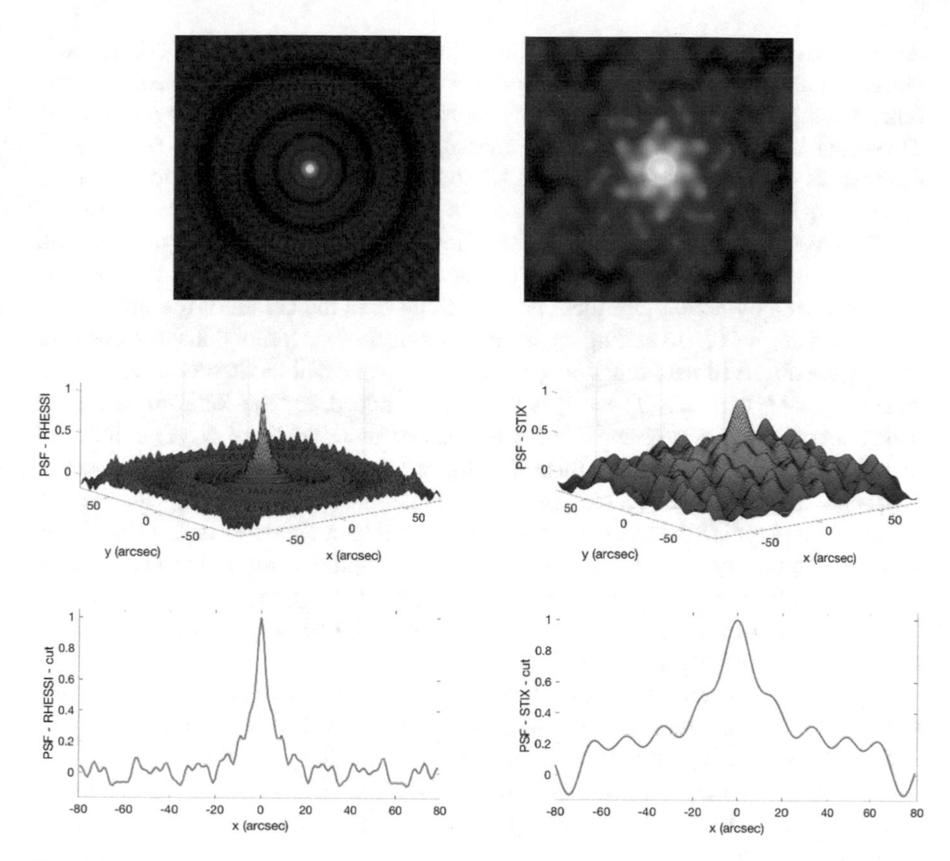

Fig. 4.2 Left column: the *RHESSI* PSF; right column: the *STIX* PSF. First row: image-like description of the two PSFs. Second row: 3D view; Third row: 1D cut passing through the center along the x axis.

to be the most concerning issue. In order to explain this last statement, consider a somewhat formal description of the image reconstruction problem, expressed by the equation

$$g = A(I) , \tag{4.8}$$

where g represents the experimental dataset provided by a spacecraft-borne instrument (e.g., for *RHESSI*, a discrete set of angular Fourier transforms [i.e., visibilities] $V(u, v)$), I is the image that we desire to reconstruct from the data, and A mimics the image formation process. For example, in the case of *RHESSI*, A is the Fourier transform sampled at the finite set of points in the spatial frequency plane corresponding to the various orientations of the nine modulation grids.

The mathematical spaces to which I and g belong are totally different: the (physical) source function I belongs to a space X consisting of functions that encode the physical properties of the X-ray emitting source, while the data function g contains not only information pertinent to I, but also additional stochastic components representing statistical properties of the noise affecting the measurement process. These noise components project the measured dataset g out of the *range* of A (defined as $R(A) = \{A(I) \ , \ I \in X\}$), so that (1) it is impossible to find I by solving Eq. (4.8) directly, and (2) *for a given g, a solution I may not even exist.*

The properties of the operator A also allow characterization of the other issues, namely uniqueness and stability. For example, the non-uniqueness of the solution is readily seen by noting that there are functions I_o in the kernel of the operator A such that $A(I_o) = 0$, and adding an arbitrary number of such functions to I does not change the observed data function g at all. This means that the kernel of A defined as $N(A) = \{I \in X \ , \ A(I) = 0\}$ is not trivial; indeed, for the *RHESSI* and *STIX* image reconstruction problems, there is a set of null functions $N(A)$ comprised of *all* the images that can be formed using arbitrarily assumed visibility values $V(u, v)$ at *all* spatial frequencies (u, v) other than those actually sampled. Finally, we note that A typically has the property of smoothing out features in I, so that large variations in the source function I can produce negligible changes in the dataset g; this implies, from a mathematical perspective, that A^{-1} is unbounded, i.e., that any image reconstruction process must address issues of numerical instability related to the presence of noise in the measured data.

Despite this apparently chaotic situation, there is hope. Most problems associated with both noise in the data and with the sparsity of Fourier sampling can be addressed by applying an a priori assumption of smoothness in the image $I(x, y)$. This is the basis for *regularization* techniques (see Sect. 4.3), which exploit plausible characteristics of the source function in order to suppress artifacts due to data noise and/or finite sampling.

4.3 The Regularization Concept

With the seminal papers of Miller [140] and Tikhonov [187], mathematics found the way to account for the pathologies of linear ill-posed problems by formulating a very general framework which currently represents the source of most image reconstruction algorithms. Regularization theory relies on a basic mathematical principle: a reliable approximate solution of an ill-posed problem can be obtained by shrinking the solution space via the introduction of constraints that encode a priori knowledge on the solution itself. This constraining procedure can be explained according to two different perspectives.

Deterministic regularization is the approach followed by Tikhonov and Miller when they introduced the regularization concept. Their basic idea was to give up looking for solutions of Eq. (4.8) but rather to compute an approximate solution by addressing the constrained problem

$$\begin{cases} \|A(I) - g\|_Y \leq \epsilon \\ C(I) \leq \mathcal{E} \end{cases} . \tag{4.9}$$

Accordingly, the goal becomes to determine an image solution I that reliably fits the data g within a normalized bound ϵ characterized by the type of norm used Y (first inequality in Eq. (4.9)) and, at the same time, satisfies a specific constraint that encodes some a priori information on the solution I, represented by imposing a bound $\mathcal{E}$ on a prescribed functional C of I (second inequality in Eq. (4.9)). Examples of such constraints are that the global energy associated with the imaging signal must be bounded, or that I be non-negative, or that I have some degree of smoothness. As applied to the image reconstruction problem, imposing a bound on the energy of the global solution implies numerical stability and so reduces image artifacts due to unphysical oscillations in the pixel content. Imposing a non-negativity constraint $I \geq 0$ simply means that the content of each pixel in the reconstructed image cannot be negative since physically I is related to the intensity of radiation coming from the Sun.

The second approach to regularization is *statistical regularization*, which employs a Bayesian approach. Here the goal is no longer to reconstruct the *"best"* approximation of the image, but rather to maximize the degree of knowledge we have about the image. The main assumptions of the Bayesian approach are that all variables involved in the problem are random variables and that the corresponding probability densities encode the degree of knowledge about such variables. The Bayes theorem [17]

$$p(I|g) = \frac{p_{pr}(I)\, p(g|I)}{p(g)} \tag{4.10}$$

connects these densities. The likelihood distribution $p(g|I)$ is based on two models: the direct problem (4.8) and the statistical model of the noise affecting the measured data g. The *prior density* $p_{pr}(I)$ encodes all a priori information we have on the image to reconstruct, and therefore is the essence of the regularization process; this information may be concerned with either physical or mathematical knowledge or may be provided by images acquired via other instrumentation operating, for example, at different wavelengths. The *posterior distribution* $p(I|g)$ is the solution of the Bayesian inverse problem; it reflects the new information we have about the image I based on accommodating the new data g. The denominator in Eq. (4.10) serves as a normalization constant.

The formal solution of the Bayesian inverse problem is the posterior distribution $p(I|g)$. However, from a practical viewpoint, $p(I|g)$ is typically a function defined in a very large mathematical space and therefore its interpretation may be complicated. Therefore, it is much more helpful to provide solutions computed by means of formulas that utilize this posterior density distribution. Two typical examples are the *conditional mean*

$$I_{cm} = \int I\, p(I|g)\, dg\ , \tag{4.11}$$

the average of all possible images I, weighted by the conditional probability of obtaining that image given the data g, and the *maximum a posteriori argument* (the image I that corresponds to the maximum in the conditional probability $p(I|g)$):

$$I_{map} = \arg\max_{I} p(I|g)\ . \tag{4.12}$$

The computation of the conditional mean involves numerical integration that can introduce further computational issues; nevertheless the conditional mean is always well defined. On the contrary, the computation of the maximum a posteriori argument is an optimization problem whose solution may not be unique (and may even not exist).

The Bayesian approach can be extended to the solution of dynamical (or spectral imaging) problems. In these cases observations are made at several time points (or in several energy channels) and the corresponding images must be reconstructed at the corresponding time points (energy channels). Formally, the measured data compose a *stochastic process* $\{g_k\}$, where $k = 1, \ldots$ indicate these points or channels and, analogously, one has to reconstruct the stochastic process of images $\{I_k\}$, $k = 1, \ldots$, given specific assumptions (the so-called Markov hypotheses). The computational solution process is the sequential scheme

$$p(I_k|g_1, \ldots, g_k) = \frac{p(g_k|I_k)\, p_{pr}(I_k|g_1, \ldots, g_{k-1})}{p(g_k|g_1, \ldots, g_{k-1})}\ , \tag{4.13}$$

$$p(I_{k+1}|g_1, \ldots, g_k) = \int p(I_{k+1}|I_k)\, p(I_k|g_1, \ldots, g_k)\, dI_k\ , \tag{4.14}$$

where Eq. (4.13) is the Bayes theorem for the computation of the posterior distribution corresponding to I_k and Eq. (4.13) is the Kolmogorov-Chapman equation, which allows computation of the prior distribution at a given time (or energy channel), $k + 1$, by using the posterior distribution provided by the Bayes theorem and a transition kernel that encodes a priori dynamical (or spectral) information. A clever choice for this kernel allows the filtering procedure to account for a smooth behavior of the source's characteristics along the temporal (or spectral) axis.

As a concluding remark, we notice that regularization methods always involve input parameters that must be optimized according to some specific criteria. In the context of the current work, the main issue is that while constraining the source space where the solution of the image reconstruction problem (4.8) lies fosters stability of the reconstruction, it necessarily also worsens the ability of the reconstructed image to accurately fit the measurements (i.e., the fidelity of the recovered image). Therefore, all regularization methods require a balance between the *stability* and *fidelity* of the solution; this process is typically realized by the optimal selection of some input parameters that are part of the reconstruction scheme. Examples of optimization algorithms that have been formulated with this objective include the discrepancy principle [188], Generalized Cross Validation (GCV) [80], and the L-curve method [84].

4.4 Numerical Optimization

The implementation of regularization methods in effective codes for image reconstruction always requires, at some stage, the numerical solution of an optimization problem. This is the case, for example, every time one has to address either constrained problems like (4.9) or maximization problems like (4.12). And this is particularly crucial in forward-fitting approaches, whereby the reconstructed image is a parametric form modeled by a set of imaging parameters that must be computed by comparison with χ^2 values.

More formally, let us use Θ to denote the set of such parameters belonging to the feasibility region $\mathcal{D}$, $I(\Theta)$ to denote the parametric model of the emitting source, and $\chi^2(\Theta)$ to denote the square of the discrepancy between the empirical data and the one predicted by the data formation model $A(I(\Theta))$. The forward-fitting approach requires the solution of the optimization problem

$$\arg\min_{\Theta \in \mathcal{D}} \chi^2(\Theta) , \tag{4.15}$$

which can be addressed by means of both deterministic and statistical techniques. In a deterministic framework, for example, VIS_FWDFIT in the HESSI GUI within SSW implements a simplex algorithm as described in Sect. 6.3. On the other hand, a Sequential Monte Carlo (SMC) approach will be described in Sect. 6.4, while a Particle Swarm Optimization algorithm [41] has been recently utilized in the case of semi-calibrated visibilities measured by *STIX* in November 2020 [16].

Chapter 5
Count-Based Imaging Methods

Abstract Analysis of the *RHESSI* and *STIX* data requires some novel tools, all tailored to, and indeed in many cases optimized for, the construction of an image from a sparse set of relatively noisy Fourier components obtained using either the temporally modulated count rates measured with *RHESSI* or the Moiré patterns measured with *STIX*. The raw data from both instruments used for all of the image reconstruction algorithms to be discussed in this chapter consist of count rates accumulated into a relatively large number ($\sim 10^3 - 10^6$) of short time bins of $\sim$0.5–100 ms each. The essence of many of these image reconstruction methods has been described by Hurford et al. (Solar Phys 210:61–86, 2002) for *RHESSI* and by Massa et al. (Astron Astrophys 624:A130, 2019) for *STIX*. Here we discuss several of them in some detail, and we add a discussion of methods that have been developed more recently. Example applications using solar flare data are presented.

5.1 Back-Projection

The back-projection method [96] essentially takes each detected count and forms a map showing the set of locations in the field of view where the responsible photon could have originated. In the *RHESSI* framework, because the photon must pass through both of two parallel grid pairs in order to be detected, the map produced by this process is simply a set of stripes across the solar disk oriented parallel to the grid slits, with the orientation and separation of the stripes dependent on the grid pitch and orientation at the time of observation, for the detector that recorded that count. When consideration is taken of partial transmission (e.g., at high energies or near the edge of a grid slit), these maps are better termed "probability maps." As the spacecraft rotates, the orientation of these stripes rotates about the spin axis, and adding the various count probability maps provides a "maximum likelihood" image of the source. This is accomplished in practice by taking a prescribed time interval and chosen range of count energies, and weighting the nine back-projection maps (one for each detector) at the center of that time interval by the number of counts observed by each detector in that time interval and in the prescribed energy range.

M. Piana et al., *Hard X-Ray Imaging of Solar Flares*,
https://doi.org/10.1007/978-3-030-87277-9_5

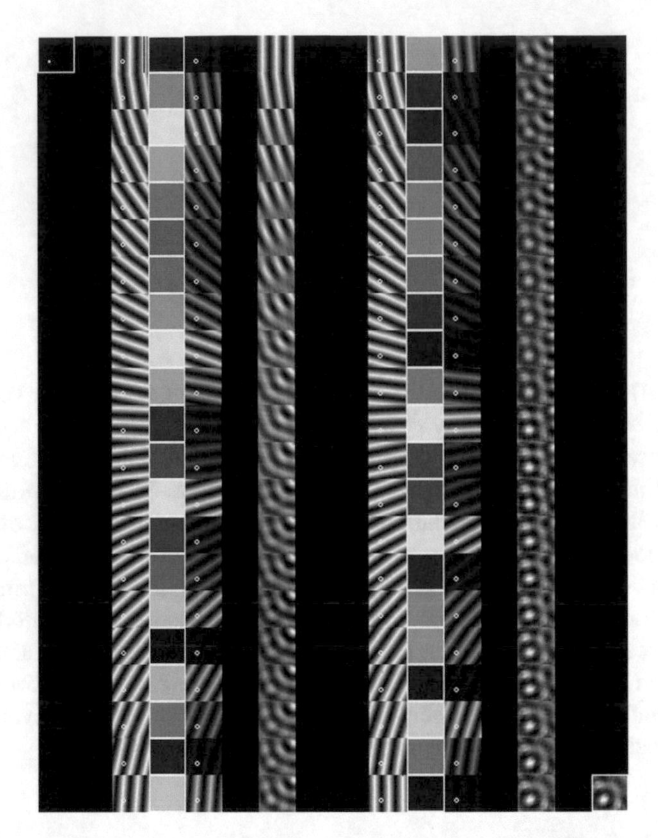

Fig. 5.1 The left and right strips in this figure show the back-projection image being built up as the spacecraft turns through a full rotation, for a grid pair with a pitch of $10''.5$. The simulated point source is shown in the top left, and also as a small circle on each image. The final reconstructed image is shown in the bottom right. Each of the two column groups shows 21 time intervals during one half-rotation of the spacecraft ($\simeq$2 s), with the four images in each row showing the following: (1) the probability map of where a single photon could have originated on the Sun in that time interval (white = highest probability), with the stripes resulting from the alternating RMC transmission and absorption of the photon depending on its point of origin, (2) the measured count rate of the detector behind the RMC in that time interval, (3) the probability map of column 1, weighted by the count rate of column 2, and (4) the accumulated sum of the weighted probability distributions in column 3. A back-projection image for longer time intervals is obtained by continuing the accumulation process for an integer number of complete rotations.

Each of these elementary back-projection maps essentially corresponds to the intensity of a two-dimensional spatial Fourier component corresponding to the wavevector **k**, where the direction of **k** is perpendicular to the grid pair used and the magnitude of **k** corresponds to the (inverse) angular resolution of the grid pair in question. As shown in Fig. 5.1, as the spacecraft rotates the continued superposition of an increasing number of these elementary back-projection maps results in progressively more complete information on the image.

As pointed out in [96], "flat-fielding [adjusting the gain at each image point to ensure that a uniform source produces a uniform image] is necessary since the sensitivity [at each point in the image] is proportional to the variance of the modulation profile for each map point." In practice, a modified version of the flat-fielding correction and normalization recipe described[1] in [56] is applied with other more detailed correction factors to ensure that the expectation value at the peak of the back-projection map equals the strength of a dominant source in the image and that the map has units of counts cm^{-2} s^{-1}.

Another consideration in making an optimum back-projection image is the weighting of the contributions of the different detectors. There are currently two weighting schemes implemented in the *RHESSI* software: *natural* weighting (in which all collimators have equal weight), and *uniform* weighting (in which the flux measured in each collimator is weighted by the inverse of its FWHM). On one hand, natural weighting optimizes the image sensitivity and so is preferred for source *detection*, but it also increases the contribution of sidelobes from extended sources to the back-projection map, thus degrading the quality of the compact source images obtained. On the other hand, uniform weighting preferentially weights the high-frequency components; it reduces the contribution of sidelobes at large spatial scales and so provides greater image quality for compact sources. For imaging of larger extended sources, natural weighting is a better option.

It is also possible, for both natural and uniform weighting schemes, to introduce a variable tapering of detector weights such that the weight for the ith detector is multiplied by the following additional factor:

$$wt_i = \exp\left[-0.89\left(\frac{\text{taper}}{\text{FWHM}_i}\right)^2\right], \tag{5.1}$$

where "taper" is a parameter chosen by the observer. Using a non-zero value for the taper parameter has the effect of gradually suppressing information from the lowest-angular resolution grids (see Fig. 5.2), thereby smoothing the image to avoid over-resolution and improving the overall PSF. For taper values less than about $4''$, the additional detector weighting due to tapering is negligibly different from unity for detectors #s 5 (FWHM $\simeq 20''$) through 9 (FWHM $\simeq 180''$). In practice, uniform weighting is generally used with the taper set to zero. If there is no evidence of modulation for the finer grids, they can be removed from the analysis, since they would only add noise to the reconstructed image.

With noise-free data and an unlimited number of imaging grids (and hence spatial Fourier components), the back-projection method would produce an accurate, high-fidelity image. In practice, however, because of both Poisson noise and the limited number of grid orientations and pitches (i.e., Fourier components) available, the back-projection process results in a so-called "dirty" map, contaminated by spurious

[1] See https://hesperia.gsfc.nasa.gov/rhessidatacenter/imaging/BPClean_changes.pdf for more details.

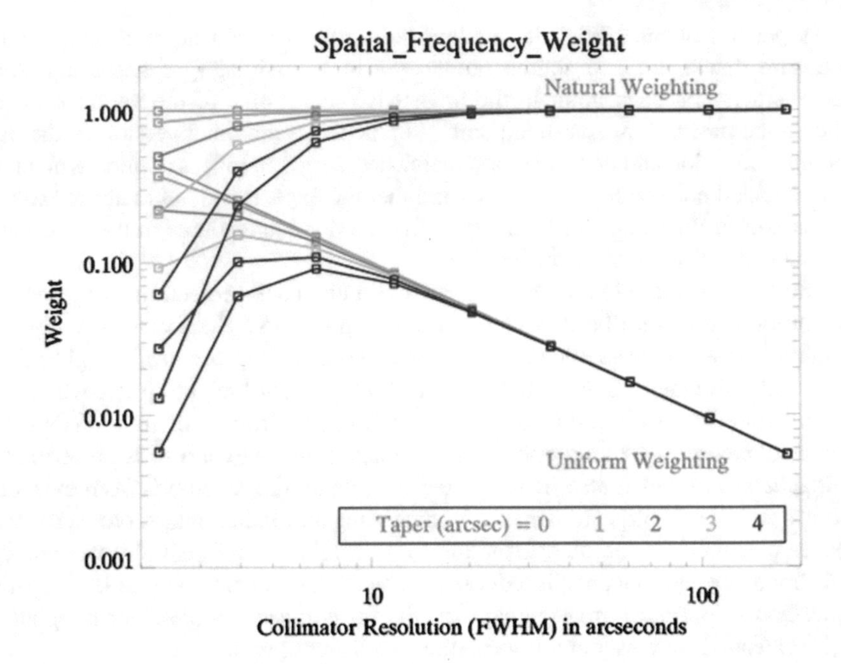

Fig. 5.2 Natural and uniform weighting schemes available in the *RHESSI* software with four different taper factors (Eq. (5.1)) shown.

features resulting from uncancelled side lobes in the limited (and sparse) set of two-dimensional spatial Fourier components sampled. Nevertheless, the back-projection map represents a starting ground-truth for the application of many of the more sophisticated image reconstruction algorithms, each aimed at extracting the true image starting from this "dirty" map.

5.2 CLEAN

The CLEAN method of image reconstruction considers the image as being comprised of multiple point sources. It was originally designed [90] to improve the "dirty" map obtained from radio interferometric observations of astronomical sources.

The core of the CLEAN algorithm is the so-called *CLEAN loop*, within which the brightest pixel in the back-projection map is successively identified in each iteration. For each such brightest pixel, a user-selected fraction or "gain" (typically $\sim$5%) of the intensity at that point is added to the *CLEAN component map* at that pixel location, while an identical source, blurred by the PSF (including its side lobes), is subtracted from the back-projection map. This process is repeated for a prescribed number of iterations (typically 100) or until some criterion of image quality is reached. In practice this stopping criterion is based either on (1) the remaining back-projection map having a negative peak larger than the biggest positive peak, or (2) that differences between the count-rate modulation profile predicted from the CLEAN component map and the observed modulation profile produce an acceptable value of χ^2. In this way, counts are sequentially "moved" from a (blurred) region of the back-projection map into a corresponding set of distinct point sources on the clean components map (see Fig. 5.4). The process preserves the total intensity represented by the sum of the two maps.

As a final step, the CLEANed map is constructed by convolving the CLEAN components map with an idealized version of the instrumental PSF, here termed the *CLEAN beam*. In practice, this PSF is an approximate Gaussian function with a width determined from the combined resolutions of the specific detectors used to make the back-projection image. The optimum width of this Gaussian function is controversial, but comparisons with images made with other techniques show that a Gaussian width equal to about half the width of the default Gaussian gives the best agreement with the source dimensions. ([110] find that a factor of 1.7 reduction in the clean beam width gives source spatial dimensions similar to those produced by other algorithms.) Note that [49] omitted this last step and used the locations and intensities of the individual point sources in the CLEAN component map to obtain the source dimensions without adding the uncertainties associated with this Gaussian blurring.

Formally, CLEAN solves the convolution equation

$$g(x, y) = (K * I)(x, y) \equiv \int\int K(x - x', y - y')\, I(x', y')\, dx'\, dy' , \qquad (5.2)$$

where $I(x', y')$ is the unknown source flux map, $g(x, y)$ is the dirty map, and $K(x - x', y - y')$ is the PSF of the instrument (see Eq. (4.1) and Sect. 4.1.2). As illustrated in Fig. 5.3, CLEAN models the solution $I(x', y')$ as the sum of Q point sources, each described by a two-dimensional Dirac δ-function, plus a background term, i.e.,

$$I(x', y') = \sum_{q=1}^{Q} I_q\, \delta(x' - x_q, y' - y_q) + B(x', y') , \qquad (5.3)$$

where I_q and (x_q, y_q) are respectively the amplitude and position of the q-th source, and $B(x, y)$ models the background. Substituting Eq. (5.3) into Eq. (5.2) leads to

$$g(x, y) = \sum_{q=1}^{Q} I_q\, K(x - x_q, y - y_q) + (K * B)(x, y) . \qquad (5.4)$$

Given an estimate of the dirty map $g(x, y)$ and knowledge of the instrument PSF $K(x, y)$, CLEAN iteratively estimates the number of required sources Q, as well as their positions and intensities $(I_q; x_q, y_q)$; $q = 1, \ldots, Q$.

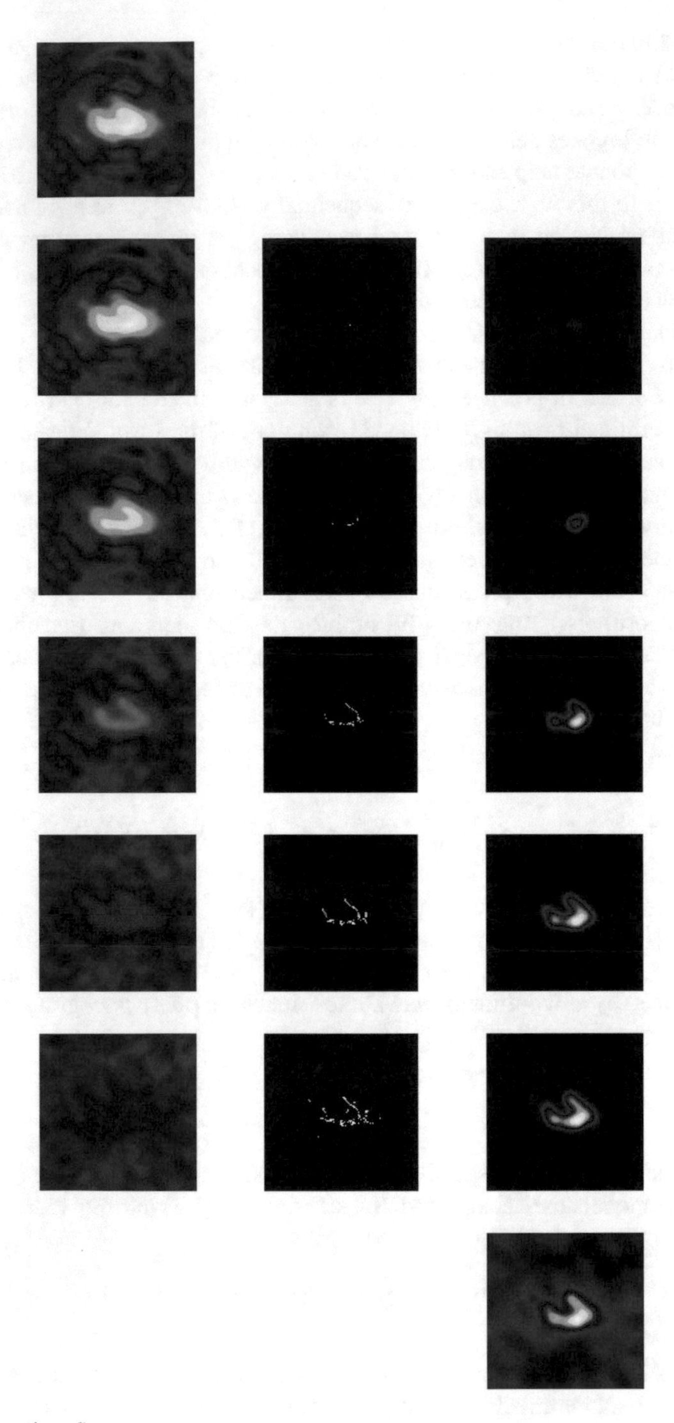

Fig. 5.3 (continued)

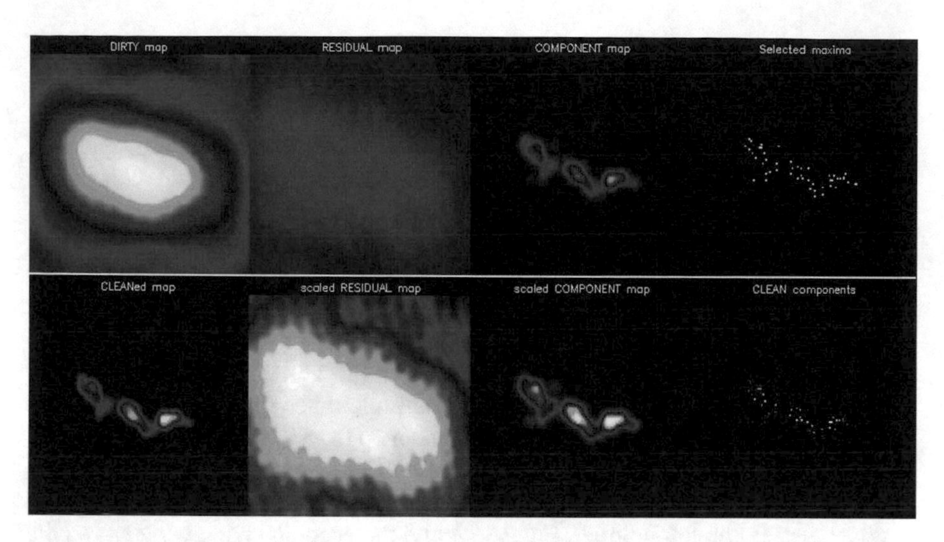

Fig. 5.4 Images showing various maps produced in the application of the CLEAN process to a 12–25 keV image during a flare on 2002 March 17 from 19:27:40 to 19:28:56 UT. The eight images are described in the table below.

DIRTY map	RESIDUAL map	COMPONENT map	Selected maxima
Starting back-projection or dirty map	Remaining map after all identified CLEAN components and their side lobes have been removed. (Scaled to the dirty map peak)	CLEAN component point sources convolved with clean beam. (Scaled to the dirty map peak)	CLEAN point-source component locations
CLEANed map	Scaled residuals	Scaled components	CLEAN components
Final CLEAN image equal to sum of component and residual maps	Residual map scaled to the peak value	Component map scaled to the peak value	CLEAN components scaled to the summed flux in each pixel

Fig. 5.3 (continued) Images illustrating the different stages of the CLEAN process after 1, 10, 50, 100 and 164 iterations (see text for details). The left column shows the original "dirty" map at the top, with the residual images after each stage of iteration shown below. The center column shows the cumulative locations of the point sources identified as the brightest pixels in the residual images. The right column shows the convolution of the images in the center column with the "CLEAN beam." The final image is the bottom image in the right column made up of the sum of the last image immediately above it and the last residual image in the left column.

Figure 5.3 illustrates the steps in, and results of, the CLEANing process. The procedure starts with the "dirty" map produced by the back-projection process described in Sect. 5.1. This is shown as the top image in the left column. The brightest pixel in this map is identified, its flux scaled by a gain factor of typically 5%, and then convolved with the instrument PSF that includes the side lobes. The resulting feature is subtracted from the dirty map leaving a residual map that is used for the next iteration. The scaled flux from that pixel is added to the corresponding pixel of a map of the so-called CLEAN point-source components (shown in the center column). The images in the right-hand column are the clean component maps convolved with a circular Gaussian (called the "clean beam") with a FWHM chosen to approximate the weighted resolution of all the RMCs used to make the original dirty map. This process is repeated multiple times (typically 100 or more) by identifying the brightest pixel in the residual map from the previous iteration and performing the same steps as before. The final CLEAN image shown at the bottom of the right column is the sum of the last image immediately above it and the last residual map in the left column. Figure 5.4 and the accompanying table shows the various maps produced in the CLEAN process applied to a 12–25 keV image of a flare on 2002 March 17 (19:27:40–19:28:56 UT).

5.2.1 Two-Step CLEAN Method

As discussed by Hurford et al. [96], *RHESSI* imaging spectroscopy is strongly limited by dynamic range, the ratio of the brightness of the strongest source in an image to that of the weakest source that can be detected in that image. The major limitation on the achievable dynamic range is in the ability to accurately determine the side lobes from the dominant source(s) against which the weaker sources must be detected. The following factors determine this capability:

- Because of its Fourier-based imaging technique, *RHESSI* necessarily includes all sources in the field of view (i.e., the full Sun). The Fourier components of the emission from weaker sources are often swamped by emission at the same spatial frequencies as from stronger sources.
- The ability to detect weak sources is limited by the Poisson statistics of the number of recorded counts in the selected time and energy bins.
- The ability to detect subtle features is limited by imperfections in the grids and by systematic uncertainties in their measured characteristics and relative alignments.

The design goal for *RHESSI* was that in the most favorable cases with high counting rates, a dynamic range of 100:1 could be achieved. In practice, dynamic ranges of 10:1 are readily achieved by most image reconstruction algorithms and 50:1 has been achieved in larger events.

An exception to this dynamic range limitation is the case where the brightest source is compact, covering only a few arcseconds in extent, with a very high surface brightness. In that case, a weaker but more extended source can produce the bulk of the emission at low spatial frequencies, and so may be detected with a smaller surface brightness than would be expected from the limited dynamic range, *as long as the image reconstruction algorithm recognizes that all the counts in a given region belong to the same source*. This is not possible with the standard CLEAN algorithm since it assumes that all sources are made up of multiple independent point sources. The Pixon algorithm (Sect. 5.4) handles this situation by considering sources with different spatial extents to best fit the measured count rate modulations. As we will discuss below, forward-fitting algorithms (Sect. 5.3) are particularly useful in this situation, since they can specify the spatial scale for each assumed source.

Within the CLEAN algorithm, the dynamic range problem is alleviated to some extent by using the so-called *two-step CLEAN* method [118]. After all compact sources are judged to have been extracted from the back-projection map and included in the CLEANed image, constructed in the usual way with all available detectors, further cleaning is done with only those detectors behind the coarser (lower spatial frequency) grids. In this way, the more extended sources can be imaged largely free from contamination from the uncleaned side lobes of the more compact sources, which make a negligible contribution to the power at small spatial frequencies. This ability to detect extended sources in the presence of compact sources is particularly relevant for the detection of *albedo sources*, sources produced by reflection of downward-propagating primary source photons off the solar photosphere [15, 102, 106]. It has been pointed out [100] that the VIS_FWDFIT method (Sect. 6.3) can better account for the low intensity and diffusiveness of the albedo component than other imaging methods currently available.

5.3 Forward Fit

The forward-fit methodology [7, 96] assumes a parametric functional form for the source (e.g., a single two-dimensional circular, elliptical or curved Gaussian, or multiple such sources), and determines the best-fit parameters by comparing the modulated count rate lightcurves predicted from the assumed source or sources with those observed.

The original forward-fit algorithm [7] for analyzing *RHESSI* (and *STIX*) observations starts with the assumption that there is a limited number of individual sources, usually only one or two. The parameters describing each source (e.g., the flux, location, and FWHM size for a circular Gaussian, plus the eccentricity and the orientation of the major axis for an elliptical Gaussian) are adjusted until the model-predicted count rates agree best (in a χ^2 sense) with the measured count rates. The photometric accuracy of this image reconstruction method, together with other available methods, was evaluated in [8].

The forward-fit method was used in [6] to analyze the flare of 2002 February 20, an event that is the subject of various image reconstruction methodologies discussed in other parts of this book, including Pixon (Sect. 5.4), uv_smooth (Sect. 6.6), and

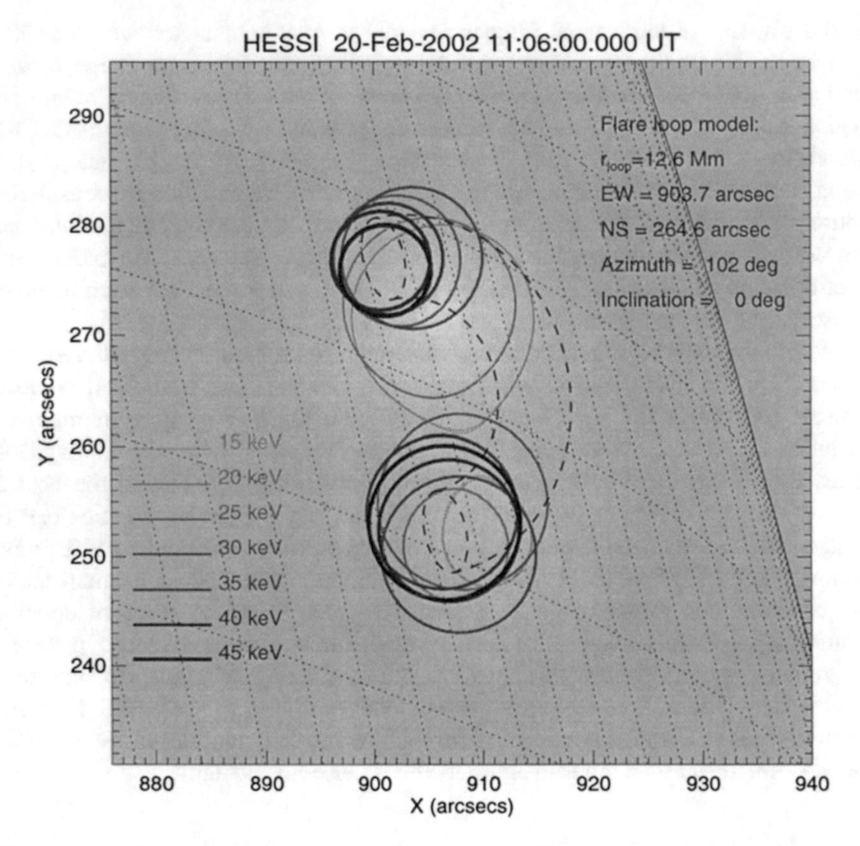

Fig. 5.5 *RHESSI* observations of the 2002 February 20 event, over the time interval from 11:06:00 UT through 11:06:40 UT. Images were made using the forward-fit algorithm assuming two independent Gaussian sources, for a variety of different energy bands. The indicated locations of the 50% contours of each Gaussian show a systematic tendency for the sources at higher photon energies to be displaced to the left which, given the geometry of the flare, corresponds to greater depths. After [6], used with permission from Springer.

Bayesian inference (Sect. 7.1). The forward-fit study in [6] assumed that there were two circular Gaussian sources with their 50% contours shown in Fig. 5.5, for seven different 5 keV energy bands from 15–45 keV. The images were made using *RHESSI* detectors #3 through 9 over a 40 s time interval from 11:06:00 UT to 11:06:40 UT. As is clear from the figure, this event occurred near the Western limb[2] of the Sun. Such a viewing angle means that differences in vertical position along the loop near the chromospheric footpoints appear as mostly lateral displacements in the plane

[2] Note that, in viewing the solar disk from the Earth, solar North is at the top, but the East→West motion of the Sun in the sky is from left to right, so that displacements to the right of the disk are Westward and those to the left are Eastward, contrary to one's immediate intuition. In a sense, we are viewing the compass rose "from the inside."

of the sky, with greater depths corresponding to points farther to the East (i.e., to the left in Fig. 5.5). Thus, it is apparent from the 50% contours shown in Fig. 5.5 that the Northern footpoint source at higher energies is located at greater depths than at lower energies. Such a behavior is consistent with the greater penetration of higher energy electrons and can therefore be used to infer the density structure of the flaring chromosphere, as we will now show. Solving the energy loss rate Eq. (1.17) in a collisional plasma gives the electron energy E as a function of column depth $N = \int n(z)\, dz$ (cm^{-2}):

$$E = \sqrt{E_o^2 - 2KN} \; , \tag{5.5}$$

where E_o is the injected energy. Continuity of the electron flux $F(E)$ (electrons cm^{-2} s^{-1} keV^{-1}) requires that

$$F(E)\, dE = F_o(E_o)\, dE_o \;\; , \tag{5.6}$$

which, combined with Eq. (5.5), yields the variation of electron flux $F(E)$ with column density N:

$$F(E, N) = \frac{E}{\sqrt{E^2 + 2KN}} \, F_o(\sqrt{E^2 + 2KN}) \; . \tag{5.7}$$

This (cf. Eq. (1.11)) leads to an expression for the variation of the observed hard X-ray intensity with height

$$I(\epsilon; z) \propto \int_\epsilon^\infty \frac{E}{\sqrt{E^2 + 2KN(z)}} \, F_o(\sqrt{E^2 + 2KN(z)}) \; Q(\epsilon, E)\, dE \; , \tag{5.8}$$

where $Q(\epsilon, E)$ is the bremsstrahlung cross-section (in their analysis, [28] used the Kramers cross-section (1.9)). Applying this expression, with a power-law injected spectrum $F_o(E_o) \sim E_o^{-\delta}$, to the observed positions of the source centroids at different energies, [6] were able to derive (actually, forward-fit a power-law form to) $N(z)$ and hence, through the relation $n(z) = dN(z)/dz$, the density vs. height $n(z)$ profile for the flaring chromosphere was determined. The resulting $n(z)$ structures (Figure 5 in [6]) compared favorably with other empirical models of the quiet Sun [196], although perhaps a better comparison would have been with empirical models of the active region chromosphere [127].

5.4 Pixon

5.4.1 Maximum Entropy Methods

Before we explore the details of the Pixon image reconstruction method, it will be instructive to first briefly examine the underlying concept of a maximum entropy method (MEM), first introduced in the context of statistical thermodynamics by E. T. Jaynes [99]. Suppose we have a set of N objects, to be distributed among n possible locations or "bins," such that the number of objects in each bin is N_i, $i = 1, \cdots n$, and $\sum_{i=1}^{n} N_i = N$. The probability of a particular "macrostate," i.e., specified set of bin counts $\{N_i\}$, is given by standard combinatorial theory as the number of outcomes that result in such a macrostate, divided by the total number of possible ways ("microstates") in which N objects can be distributed among the various bins:

$$P(\{N_i\}) = \frac{N!}{n^N \prod_{i=1}^{n} N_i!} . \tag{5.9}$$

If the count numbers N_i are substantial, we may use Stirling's formula $\ln N! = N \ln N$ to approximate the factorials in Eq. (5.9). Introducing the individual bin probabilities $p_i = N_i/N$, we obtain, for the overall probability of a prescribed macrostate,

$$\begin{aligned} \ln P(\{N_i\}) &\simeq N \ln N - N \ln n - \sum_{i=1}^{n} N\, p_i \ln (N p_i) \\ &= N \ln N - N \ln n - \sum_{i=1}^{n} N \ln N\, p_i - \sum_{i=1}^{n} N\, p_i \ln p_i \\ &= N \ln N \left(1 - \sum_{i=1}^{n} p_i\right) - N \sum_{i=1}^{n} p_i \ln p_i - N \ln n \\ &= -N \sum_{i=1}^{n} p_i \ln p_i - N \ln n , \end{aligned} \tag{5.10}$$

since $\sum_{i=1}^{n} p_i = 1$. Recognizing that $N \ln n$ is a constant, the overall probability of the set of outcomes $\{N_i\}$ is therefore maximized when the quantity

$$S = -\sum_{i=1}^{n} p_i \ln p_i \tag{5.11}$$

is a maximum. Very similar arguments are is used in statistical thermodynamics[3] [e.g., 189], where the quantity S is (within a factor of Boltzmann's constant k) the thermodynamic *entropy* of the system.

Maximizing the value of the entropy function S must be balanced with ensuring that the $p_i = N_i/N$ values are as consistent as possible with the observations; in practice this is done by maximizing the function

$$S - \lambda\chi^2 , \tag{5.12}$$

where χ^2 is a measure of the goodness of fit of the model to the observations (e.g., the modulation profiles in the various detectors, or the visibility values $V(u, v)$—see Sect. 6.5) and λ is a "regularization parameter" (Sect. 4.1) that controls the balance between maximizing the entropy function S and minimizing χ^2, i.e., maintaining fidelity with the observations. In the most basic application of the maximum entropy method to image reconstruction, the count numbers N_i are simply the counts in the n individual pixels that constitute the source. To a limited extent, such Maximum Entropy methods have been used to analyze Fourier-based hard X-ray imaging data; for example, this method was used in [164] to analyze data from the HXT instrument on Yohkoh (Sect. 1.3). However, the availability of the more general Pixon method, next to be discussed, has effectively superseded methods based on pixel-by-pixel counts.

5.4.2 The Pixon Method

The Pixon method is another technique which removes the sidelobe pattern of a telescope while mitigating the problems of correlated residuals and spurious sources commonly seen in Fourier deconvolution, chi-square fitting, and Maximum Entropy (e.g., [160]). It has been successfully applied to data from the Yohkoh/HXT instrument [139] and the HXT algorithm was adapted[4] for use by *RHESSI*.

The Pixon approach [155] attempts to minimize image complexity by locally smoothing the model source as much as the data allow, thus reducing the number of independent elements in the image. Its essence [139] is rooted in the Bayesian expression

[3] In thermodynamics, the entropy is maximized subject to a set of specified physical constraints. In the simplest example, the p_i correspond to different energies E_i, and conservation of total energy requires $U = \sum p_i E_i = \text{constant}$, so that the variation $dU = \sum E_i dp_i = 0$. The entropy is a maximum when $dS = -\sum(\ln p_i + 1)\, dp_i = 0$ and, since $\sum dp_i = 0$, it follows that $dS^* = -\sum(\ln p_i + \alpha)\, dp_i = 0$, where α is a Lagrange multiplier. Writing $dS^* - \beta\, dU = 0$, where β is another Lagrange multiplier, gives $\sum(\ln p_i + \alpha + \beta E_i)\, dp_i = 0$ and, since the individual variations dp_i are arbitrary, each term in this sum must equal zero. This immediately gives the well-known Maxwell-Boltzmann distribution of energies $p_i \propto e^{-\beta E_i}$—cf. Eq. (1.1).

[4] https://hesperia.gsfc.nasa.gov/rhessi/software/imaging-software/pixon/index.html.

$$p(I, M|g) = \frac{p(g|I, M)\, p(I|M)\, p(M)}{p(g)} \propto p(g|I, M)\, p(I|M)\,, \tag{5.13}$$

where g is the data, I is the reconstructed image, M is a model structure (e.g., a sum of circular Gaussian sources of various sizes, or a sum of loop-like shapes of various sizes and aspect ratios), and we have ignored the (constant) prior probabilities $p(D)$ and $p(M)$. Both the model M and the image I are varied to produce the maximum conditional probability $p(I, M|g)$ of a reconstructed image I, given a model M and an observed data set g.

The right side of Eq. (5.13) consists of two terms: $p(g|I, M)$, the probability of obtaining a given data set g given both an image I and a model M, and the *image prior* $p(I|M)$, the probability of obtaining such an image given the constraints of the underlying model (note that this factor does not depend on the data g). A suitable form for the image prior is

$$p(I|M) = \frac{N!}{n^N \prod_i N_i!}\,, \tag{5.14}$$

where n is the number of pixons used in the model, the set $\{N_i\}$ are the count rates in pixon i, and $N = \sum_i N_i$ is the total number of counts in the image.

Comparing equations (5.14) to (5.9), we see that the Pixon method is essentially a MEM, but with the addition of a local condition which defines the spatial scale required by the data at each point in the image. The method also has commonalities with forward fitting in that the Pixon algorithm uses a superposition of elements from a family of multi-resolution basis functions ("pixons") to derive and implement an optimal model with the minimum information content allowed by the data. The goal of the Pixon method is to construct an image using the simplest possible model while maintaining consistency with the data. But instead of the entropy function being based on the distribution of counts N_i within individual *pixels*, it is instead based on the distribution of counts among "pixons" (essentially "meta-pixels," each with a specified size and geometry).

The quantity $p(g|I, M)$ is a measure of the goodness of fit to the data and is commonly expressed as $e^{-\chi^2/2}$, so that the Bayesian conditional probability of an image I, based on a pixon model M and a data set g is

$$p(I, M|g) \propto \frac{1}{n^N} \exp\left(-\frac{\chi^2}{2} - \sum_{i=1}^{n} \frac{N_i}{N} \ln \frac{N_i}{N}\right)\,. \tag{5.15}$$

Essentially, then, the Pixon algorithm constructs an image I (based on a model M) by arranging the N observed counts into a set of n pixons, the size and shape of which depend on the model M used. In doing so, it seeks to minimize the number of pixons n that are needed (i.e., to minimize image complexity), while taking into account the χ^2 deviation between the derived image and the observations. Because the total number of counts N is the same in each such arrangement, the

Pixon method necessarily produces an image with the correct spatially-integrated flux (unlike, e.g., CLEAN, for which the spatially-integrated flux in the CLEANed image is only a fraction of that in the original dirty back-projection map).

In general, Pixon is a very computationally intensive method, and its applicability has gradually increased with the exponential increase in computing power over the past couple of decades. As with other reconstruction algorithms, the Pixon method can have difficulties with the most compact sources breaking up when the finest grids are used; this problem can be alleviated with various user-controlled parameters.

Most of the literature on use of the Pixon image reconstruction method to study solar flares involves a comparison of the method with other methods, such as CLEAN (Sect. 5.2), MEM_NJIT (Sect. 6.5) and Visibility Forward Fit (Sect. 6.3). In [8] the 2002 February 20 event (see also Sects. 5.3, 6.6, and 7.1) was studied, and it was found that the Pixon and CLEAN analysis methodologies produced "a robust convergence behavior and a photometric accuracy in the order of a few percent." It was also (Figure 2 of [8]) that the Pixon method revealed slightly more detailed structure than the other methods. An East-limb double footpoint event on 2004 January 6 was analyzed in [110], where it was noted that all four methods yielded "very similar results."

The most extensive use of the Pixon method has been in [198], where the same four methods—CLEAN, Pixon, VIS_FWDFIT, and MEM_NJIT—were used to produce time series of *RHESSI* hard X-ray images for 24 flares, ranging in flare classification size from GOES class C3.4 to X17.2, and involving both compact footpoints and extended coronal sources. They found that all four methods gave generally consistent results for the hard X-ray source sizes and that "the correlations are very good for the thermal sources, and somewhat lower for the footpoints." Similar to [8], it was also noted (Figure 1 in [198]) that the Pixon method revealed slightly more detailed structure than the other methods, and it was also found that the Pixon method correlates best with CLEAN in determining a variety of source properties, from the sizes and shapes of extended thermal sources (Figures 2, 3 and 5 in [198]) and footpoint areas (Figure 6 in [198]). Overall, it was concluded that "the CLEAN and Pixon methods have proven to be the most stable techniques, while Visibility Forward Fit had problems in some particular cases, probably because of complex sources." However, it was also found that in general the MEM_NJIT method "gives systematically smaller thermal source sizes" and "has a tendency to break up larger sources," so that "MEM_NJIT should be used with care when it comes to thermal coronal sources, and that it is not well-suited for larger sources" [198]. Similar conclusions have been found when comparing the MEM_NJIT and uv_smooth methods (Sect. 6.6).

5.5 Expectation Maximization

The Expectation Maximization (EM) algorithm is based on the Lucy-Richardson Maximum Likelihood [170] method for recovering an image that has been blurred by a known point response function. It is frequently used with optical astronomical telescopes to obtain the true image from the observed image. For *RHESSI*, the observables are the time-varying count rates in each of the nine detectors. The method sets up an iterative scheme which uses the deviations between the predicted count profiles (derived from a recovered image) and the observed count profiles to make changes in the image.

Formally, EM is a statistical algorithm that maximizes the probability that the data vector is a realization corresponding to a Poisson random vector g. A *Poisson distribution* is associated with a discrete (integer) variable N that measures the number of events in a certain time interval, given that the mean number of events is λ. It has the well-known form (e.g., [71])

$$P(N) = e^{-\lambda} \, \frac{\lambda^N}{N!} \, . \tag{5.16}$$

Based on this, we consider the inverse matrix problem

$$g = A\, I \, , \tag{5.17}$$

where it is assumed that g is a data N-vector of realizations of Poisson variables, I is an N-vector representing the source distribution, and A is the matrix linking the two. As explained in Sect. 4.3, the (conditional) likelihood of obtaining the value g for the data vector is

$$p(g|I) = \prod_{i=1}^{N} e^{-(AI)_i} \, \frac{(AI)_i^{g_i}}{g_i!} \, , \tag{5.18}$$

from which

$$-\log p(g|I) = -\sum_{i=1}^{N} [\, -(AI)_i + g_i \log{(AI)_i} - \log g_i! \,]. \tag{5.19}$$

Focusing only on the quantities in Eq. (5.19) that depend explicitly on the unknown vector I gives the *Kullbach-Leibler (KL) function*

$$KL(x) = \sum_{i=1}^{N} \big[\, (AI)_i - g_i \log{(AI)_i} \,\big] \, . \tag{5.20}$$

The EM method consists of finding, in a computationally efficient manner, the vector I that minimizes the KL function (5.20) for a given data vector g. The Karush-Kuhn-Tucker (KKT) theorem provides the conditions necessary for such a minimum, viz. [12]

$$\begin{cases} I_i \geq 0 \\ \lambda_i \geq 0 \\ \nabla_{I_i} KL(I) - \lambda_i = 0 \\ \lambda_i x_i = 0 \end{cases} \tag{5.21}$$

where $i = 1, \ldots, N$ and the λ_i are Lagrange multipliers.

Calculating the partial derivative of $KL(I)$ with respect to one of the I_k one obtains

$$\begin{aligned} \frac{\partial}{\partial I_k} KL(I) &= \sum_{i=1}^{N} \left[\frac{\partial}{\partial I_k} \sum_{j=1}^{N} A_{ij} I_j - \frac{g_i}{(AI)_i} \frac{\partial}{\partial I_k} \sum_{j=1}^{N} A_{ij} I_j \right] \\ &= \sum_{i=1}^{N} \left[A_{ik} - \frac{g_i}{(AI)_i} A_{ik} \right] = \sum_{i=1}^{N} A_{ki}^T \mathbf{1}_i - \sum_{i=1}^{N} A_{ki}^T \frac{g_i}{(AI)_i}, \end{aligned} \tag{5.22}$$

where $\mathbf{1}$ is an N-vector with components all equal to unity. It follows that the gradient of $KL(I)$ is the vector

$$\nabla_I KL(I) = A^T \left(\mathbf{1} - \frac{g}{AI} \right), \tag{5.23}$$

where g/AI is the vector with ith component equal to the ratio of the respective components of g_i and $(AI)_i$.

We are now ready to describe the EM method. The third and fourth of the KKT conditions (5.21) imply that

$$I_i (A^T \mathbf{1})_i = I_i \left[A^T \left(\frac{g}{AI} \right) \right]_i \tag{5.24}$$

for $i = 1, \ldots, N$, from which

$$I_i = I_i \frac{\left[A^T \left(\frac{g}{AI} \right) \right]_i}{(A^T \mathbf{1})_i} . \tag{5.25}$$

The EM method uses an iterative approach which solves equation (5.25) through a series of successive approximations for the vector I [19]:

$$I^{(k+1)} = I^{(k)} \frac{A^T \left(\frac{g}{AI^{(k)}} \right)}{A^T \mathbf{1}}; \qquad I^{(0)} = \mathbf{1} . \tag{5.26}$$

Since the matrix A^T is ill-conditioned, this iterative algorithm must be regularized by applying some form of stopping rule. A very effective stopping algorithm [19, 20] is a sort of discrepancy principle back-projected onto the image space. Specifically, given δ_i, the level of noise that corrupts the i-th component of the data, one may want to stop the algorithm at the iteration k that satisfies

$$|g_i - (AI^{(k)})_i| \approx \delta_i \ . \tag{5.27}$$

In the Poisson case under consideration, $\delta_i = \sqrt{(AI^{(k)})_i}$. Now, the difference between two successive iterates of EM is given by

$$I^{(k+1)} - I^{(k)} = \frac{I^{(k)}}{A^T 1} A^T \left(\frac{g}{AI^{(k)}} - 1 \right) . \tag{5.28}$$

Inserting Eq. (5.27) into Eq. (5.28), leads to a discrepancy principle for the difference between two successive iterates: the algorithm should be stopped when

$$(I^{(k+1)} - I^{(k)})_j^2 \approx \left(\frac{I^{(k)}}{A^T 1} \right)_j^2 \left(A_2^T \frac{1}{AI^{(k)}} \right)_j , \tag{5.29}$$

where A_2 is obtained by computing the square of each entry of A.

The essential rationale for the EM method is as follows: for *RHESSI*, the observable Poisson-variable components g_i are the (time-varying) count numbers (for the chosen time and energy intervals) in each of the nine detectors, the source vector components I_i are the intensities in each image pixel, and the matrix A represents the instrument response. Starting from a gray map (all $I_i^{(0)}$ equal to unity; Eq. (5.26)), the ratio of each observed count number g_i to the predicted count number $AI_i^{(k)}$ (with division by zero returning a zero) is used to construct a back-projection scaling map (Sect. 5.1) of the same size as the image; multiplying the map $I^{(k)}$ by this scaling map produces the next image iteration $I_i^{(k+1)}$.

Finally, examples of an application of the EM method, to three events observed by *RHESSI* in 2002, are shown in Fig. 5.6. In this figure, EM performances are compared to the ones provided by Pixon and CLEAN. In all cases [19] the C-statistic values corresponding to EM and Pixon reconstructions are smaller than the ones corresponding to CLEAN and, in general, very close to unity. EM obtains such a comparable performance with a computational burden that is almost four times smaller than that required by Pixon.

The feasibility of the EM approach has been proven also for the reconstruction of *STIX* images [130].

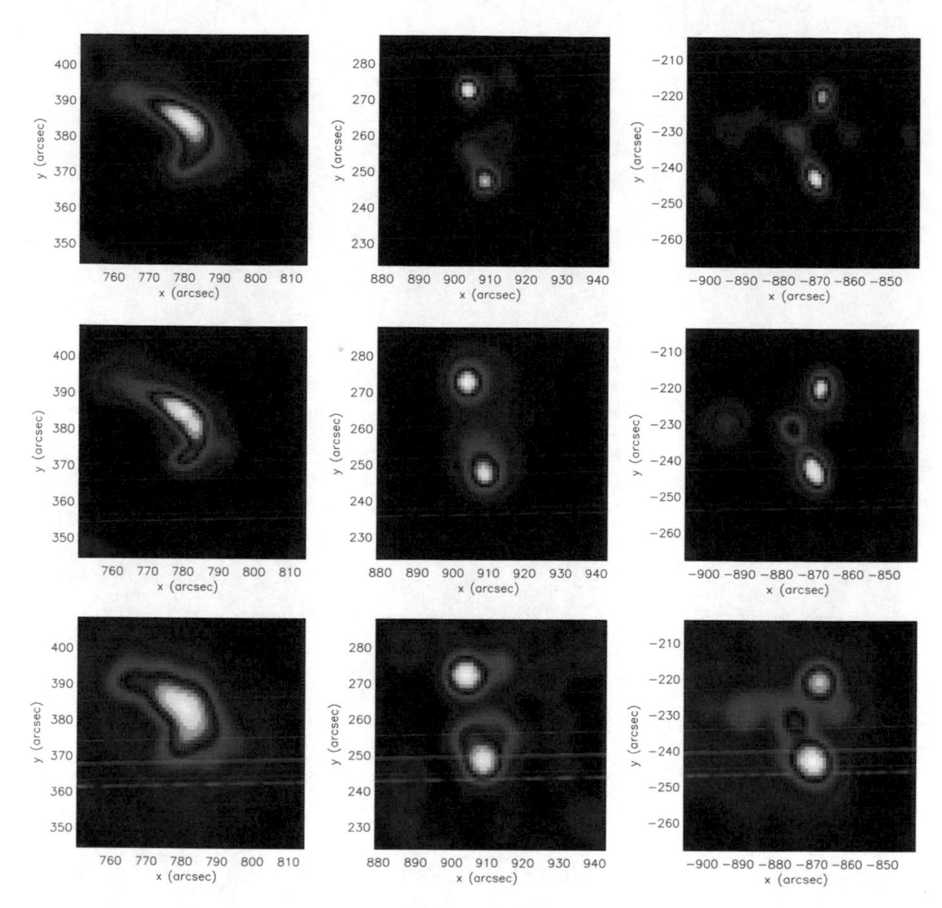

Fig. 5.6 A comparison of the performance of the Expectation Maximization (EM) method as applied to three flares observed by *RHESSI* in 2002. The first though third columns show results for the 2002 April 15, 2002 February 20, and 2002 July 23 events, respectively. *First row*: EM reconstructions; *second row*: Pixon reconstructions; *third row*: CLEAN reconstructions using the standard clean beam width (Section 5.2). After [19], reproduced with permission from the EDP Sciences for European Southern Observatory.

Chapter 6
Visibility-Based Imaging Methods

Abstract As emphasized in many places above, *RHESSI* and *STIX* are "Fourier imagers"—the native form of the data is a set of spatial Fourier components, deduced from analysis of either the temporal modulation of the detector count rates as the spacecraft rotates (in the case of *RHESSI*) or the sets of Moiré patterns (in the case of *STIX*). In this Chapter we formalize this concept through the definition of "visibilities"—components of the spatial Fourier transform of the source image at spatial frequencies sampled by the instrument. We then proceed to discuss a variety of image reconstruction algorithms that are optimized to a dataset that consists of a sparse number of measured visibilities, and we compare their strengths and limitations. Finally, we discuss an ingenious method that inverts the order of the spatial and spectral inversion processes in proceeding from count-based visibilities to images of the mean source electron spectrum, first by *spectrally* inverting the count-based visibilities to obtain the visibility values associated with the electron flux spectrum, and then performing the *spatial* Fourier-based inversion to obtain images of the mean source electron spectrum. By virtue of the manner in which they are constructed, such images vary sufficiently smoothly with electron energy E to permit application to flare studies, discussed in the next chapter.

6.1 Visibilities

In Chap. 3, we showed that the native outputs of the *RHESSI* rotating modulation collimators (RMCs) and the *STIX* fixed collimators are sets of two-dimensional spatial Fourier components $V(u, v)$ of the image, where u and v are angular spatial frequencies (with units of arcsec^{-1}) referred to suitably-oriented orthogonal x and y directions, respectively. For *RHESSI*, as the spacecraft rotates, the RMCs sample the spatial frequency (u, v) domain along nine circles with radii from R_9 $(= 1/2\alpha_9 \simeq 0.0027$ arcsec$^{-1})$ to R_1 $(= 1/2\alpha_1 \simeq 0.221$ arcsec$^{-1})$, corresponding to angular widths that increase from $\alpha_1 \simeq 2.26''$, by geometric steps of $\sqrt{3}$, to $\alpha_9 = 3^4 \times \alpha_1 \simeq 183''$. Geometric (rather than linear) steps were used since the modulation amplitude of any one collimator decreases from 90% to 10% of its maximum value

M. Piana et al., *Hard X-Ray Imaging of Solar Flares*,
https://doi.org/10.1007/978-3-030-87277-9_6

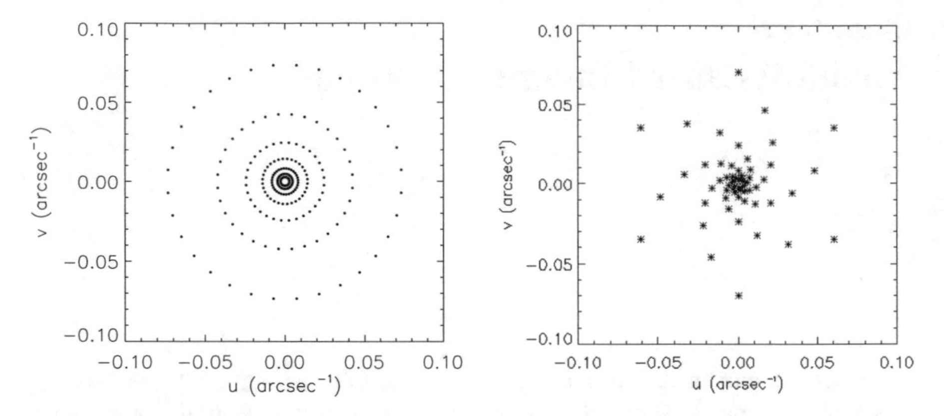

Fig. 6.1 Spatial frequency points sampled by *RHESSI* (left panel) and *STIX* (right panel). For *RHESSI*, visibilities for detectors 3 through 9 are indicated at ~32 spatial frequency points for each detector on a circle in the spatial frequency, or (u, v) plane. For *STIX*, the visibility points are fixed in the (u, v) plane at the spatial frequency points shown. After [57], used with permission from EDP Sciences for European Southern Observatory.

as the source size changes by the same *factor* of ~5. The factor $\sqrt{3}$ was chosen in order that the third harmonic of a detector n has the same spatial frequency as the fundamental mode of detector $(n-2)$, thus providing a degree of redundancy and cross-calibration—see Eq. (3.3). In the case of *STIX*, 30 sub-collimators provide 60 spatial frequency components arranged on 10 circles of the (u, v)-plane, with radii ranging from about 2.79×10^{-3} arcsec^{-1} to 7.02×10^{-2} arcsec^{-1}. Figure 6.1 shows the locations of the sampled spatial frequency components for both instruments.

The quantities $V(u, v)$ are termed *visibilities*; they have been in common use in radio astronomy for some time, since the signal from an interferometric array of radio telescopes also directly measures spatial Fourier components of the source [e.g., 186]. The idea of using visibilities in the context of *RHESSI* observations was first introduced by Hurford et al. [96], but a more detailed description was given by Hurford et al. [97].[1] Rather than constructing the image pixel-by-pixel, as is done by most conventional imaging instruments and even by the count-based image reconstruction methods described in Sect. 5.1, visibility-based methods construct the image by superposition of individual Fourier component "patterns," one pattern for each sampled (u, v) point.

A fundamental limitation of this method, which will be discussed more fully later, is that it is impossible to focus on a specific sub-domain in the field of view—*all* sources in the field of view (for both *RHESSI* and *STIX*, the whole Sun) contribute to each (u, v) Fourier component. Thus, side lobes of the Fourier pattern associated with a bright source at one location on the Sun can effectively swamp

[1] A copy of the poster presented with the abstract in [97] is available at https://hesperia.gsfc.nasa.gov/rhessi3/pics/rhessi_visibility_poster_SPD_2005.pdf.

the signal from a relatively weaker source situated elsewhere. This unavoidable feature of Fourier-transform-based imaging essentially limits the dynamic range to a factor of about 100:1; imaging of weaker sources is impossible if 100 times stronger sources are present anywhere on the Sun. Significant improvements over this basic limitation are possible for extended sources using forward-fitting techniques in both image space (Sect. 5.3) and spatial-frequency space (Sect. 6.3).[2]

There are two crucial differences between analysis of radio interferometric visibility signals and those from hard X-ray instruments:

1. First, the number of radio photons observed by an array of large radio telescope dishes is very large because both the aperture sizes are so large and the energy per photon is so small. As a result, the signal has a very low statistical noise level. By contrast, *RHESSI* and *STIX* observe a much smaller number of photons in a given range of energies because the detector areas are much smaller and the energy per photon is much larger in hard X-rays than in the radio domain.
2. Second, the number of Fourier components sampled by both *RHESSI* and *STIX* is rather small. For *RHESSI*, although the modulation profile for each collimator in principle provides visibilities measured over a continuum of orientations, in practice, obtaining sufficient counts in each time interval sampled requires that each circle in the (u, v) plane (Fig. 6.1) be sampled over a finite range of azimuthal angles ϕ_o. Thus the (u, v) plane is effectively sampled over a number of discrete arcs, each forming part of one of the circles corresponding to a given grid (Fig. 6.1). For *STIX*, the sampled (u, v) points are fixed by the number and geometry of the transmission grids (Sect. 3.6).

   For *RHESSI* the choice of azimuthal spacing is influenced by two conflicting requirements. On the one hand, the ratio of the imaging field of view to the spatial period of a given collimator defines a maximum azimuthal spacing that satisfies the Nyquist-Shannon theorem (expressed in polar coordinates) for adequate sampling in the (u, v) plane. On the other hand, a minimum spacing in azimuth is imposed by the requirement that there be adequate sampling of an intensity modulation cycle (see Sect. 3.2). This depends on the angular distance between the flare and the spacecraft spin axis. As long as that distance is $\gtrsim$300 arcsec, the optimum number of visibilities per circle[3] is $\sim$32, each corresponding to an azimuth range $\Delta\phi_o \simeq 11°$. Thus, the maximum number of independent visibilities is $\sim 9 \times 32 = 288$. In practice, the number of *statistically significant* visibility values is less than this: it may be several dozen, but seldom more than a hundred or so, even in the strongest events. By contrast, the $N(N-1)/2$ visibilities generated by a radio interferometric array of N telescopes can produce, for $N \simeq 100$, several thousand independent visibilities, all with high signal-to-noise ratios. As a result, the highly-developed visibility-based algorithms for radio

[2] See further discussion of this issue in Sect. 3.3.

[3] The default numbers of visibilities used in the IDL analysis software are as high as 64 for the detectors behind the finer grids, decreasing to as low as 6 for detector #9, depending on the flare location with respect to the spin axis.

astronomy (e.g., in the Astronomical Image Processing System (AIPS); [186]) are not necessarily well-suited to image reconstruction from *RHESSI* visibilities, and special visibility-based image reconstruction methods have been developed for use with the *RHESSI* data.

Representing the input data as a set of visibilities has several advantages. First, since each visibility is calculated from a different set of detected counts, with each measured count contributing to one and only one visibility, statistical uncertainties in the set of visibilities are independent, readily calculated (assuming Poisson statistics), and (for simple source geometries) straightforwardly propagated into statistical uncertainties in the source parameters. Second, since visibilities are linearly related to the observed count rate, they can be linearly combined and/or weighted (as a function of time or energy, for example) to suit the user's purposes. As more visibilities (i.e., spatial Fourier components) are included, reinforcement and partial cancellation of the individual Fourier component patterns in different regions of the image progressively sharpens the picture into a recognizable image. Further, visibility information does not require any background subtraction, since most background effects are not modulated by the instrument, and so do not affect the value of the inferred visibilities. However, the limited sampling of Fourier space necessarily implies that any image produced from such a sparse set of visibilities will contain a significant signal in uncancelled Fourier components, typically in the form of concentric rings (see Fig. 6.6). Finally, the signal-to-noise ratio (SNR) associated with visibilities is typically smaller than the one associated with counts [130].

We now proceed to discuss some of the visibility-based image reconstruction methods that have been developed to extract the maximum useful scientific information from a set of $\lesssim$100 sparsely distributed, and generally noisy, visibilities. In contrast to the count-based algorithms discussed in Chap. 5, visibility-based algorithms do *not* proceed with the initial creation of a "dirty" back-projection map; instead they form an image by combining the spatial Fourier components to create a sparsely-populated Fourier transform of the image which is then either forward-fit, smoothed, or inverted to produce a two-dimensional image.

6.2 Visibility-Based Methods

The following methods are employed in the various image reconstruction algorithms that are available for the analysis of *RHESSI* data.

1. Visibility-based CLEAN;
2. Visibility-based forward-fitting routines that determine the best-fit values of the parameters corresponding to assumed simple functional forms for the source by minimizing a cost function *in visibility space*;

3. Bayesian optimization as a generalization of the forward-fit approach in which the number of sources is considered as a free parameter and the uncertainties are directly determined.
4. Visibility-based maximum entropy approaches;
5. The uv_smooth method that smooths the noisy, sparsely-populated, Fourier map before attempting a transform into image space;
6. Methods that realize a decomposition of the flaring source as sparse combinations of basis functions;
7. An ingenious method that performs a spectral inversion on the visibilities in order to construct visibilities in the electron (rather than count) domain before using these "electron visibilities" to create an image of the electron flux that produces the bremsstrahlung photons observed.

We next discuss each of these methods in some detail.

6.3 VIS_FWDFIT

In Sect. 5.3, we discussed the forward-fitting of assumed, parametric, source geometries to the information contained in the count-based "dirty" back-projection map. The current version of the *visibility* forward-fit method implemented in the *RHESSI* software is called VIS_FWDFIT. As with the original forward fitting method, it also utilizes a parametric form for the assumed sources but now the forward-fitting is applied to the *visibility* data, rather than to the corresponding real-space image. In this way, the essential spatial properties of the source are derived directly from the native data, without the need to produce a noisy spatial map as an intermediate product. Also, since the uncertainties on the visibility data are known, VIS_FWDFIT provides not only the values of the source parameters, *but also their uncertainties*, using a Monte Carlo technique.

VIS_FWDFIT uses a similar approach to the original visibility forward-fit algorithm [206] in that it minimizes the difference between the measured visibilities and those expected from the assumed sources. It uses the IDL version of the AMOEBA function [173] to perform multi-dimensional minimization of the χ^2 determined from the differences between the visibilities computed from the measured counting rate modulations and the visibilities computed from the assumed sources. This approach offers fast determinations of the best-fit source parameters. However, as with any forward-fitting method, it is critical that the assumptions concerning the number of sources and their shapes and sizes closely match reality. This is particularly true for *STIX* since fewer visibilities are measured than with *RHESSI*. More effective forward-fit routines based on artificial intelligence are currently under construction in the *STIX* data analysis framework. These rely on more modern optimization approaches like evolutionary computation and deep learning.

One major drawback of the original visibility-forward-fitting algorithm [206] was the limited number and shapes of sources that could be included. This problem has been alleviated with the latest version of VIS_FWDFIT in that an unlimited number of circular, elliptical, and curved elliptical sources (shown in Fig. 6.3) can now be used. An additional capability is the ability to fix selected parameters during the fitting process, thus reducing the number of free parameters and allowing the routine to converge on a reasonable solution. For example, a source location or FWHM extent can be fixed or allowed to vary only gradually if it is expected that this would be the case over a range of times or energies. In this way, more sensitive imaging spectroscopy is possible, particularly for extended sources such as coronal hard X-ray sources in the presence of more compact and more intense footpoint sources.

The critical importance of ensuring that the assumptions concerning the number of sources, and their shapes and sizes, closely match reality is illustrated in the following example of the first application of the visibility-forward-fit method as applied to *RHESSI* data [206]. Visibilities observed in a number of flares with a simple, single-loop geometry (as determined from preliminary analysis using the CLEAN—Sect. 5.2—method) were fit with the parametric visibility function corresponding to a single curved elliptical Gaussian source. In spatial coordinates, such a source is described by the parametric equation

$$I(x, y; \epsilon) = I_o(\epsilon)\, e^{-s^2/2\sigma^2(\epsilon)}\, e^{-t^2/2\tau^2(\epsilon)}\,, \tag{6.1}$$

where $s(x, y)$ is a coordinate along a circular arc of radius R in the plane of the sky, t is the coordinate locally perpendicular to this arc (also in the plane of the sky), and $\sigma(\epsilon)$ and $\tau(\epsilon)$ are the (energy-dependent) standard deviations of the source extent in the parallel and perpendicular directions, respectively. The fit therefore involves seven parameters, each a function of photon energy ϵ: the peak intensity I_o, the location (x_o, y_o) of the center of the arc defining the source, the direction θ_o defining the orientation of the vector from (x_o, y_o) to the middle of the arc, the radius of curvature R of the arc, and the standard deviations σ and τ. The (energy-dependent) visibilities $V(u, v; \epsilon)$ corresponding to the spatial form (Eq. 6.1) can be calculated; for each ϵ they are also functions of the same seven parameters $(I_o, x_o, y_o, \theta_o, R, \sigma, \tau)$. Forward-fitting this functional form of the spatial Fourier transform of the parametric image to the measured visibilities provides the values (with uncertainties) of these seven parameters at each count energy ϵ.

Figure 6.2 shows maps of the ten simple-geometry flare events studied in [206]. The first row shows the image in the 10–15 keV energy range produced using the CLEAN method (Sect. 5.2), while the second row shows the image recovered using the visibility-forward-fit method (VIS_FWDFIT). The third and fourth rows show the CLEAN and VIS_FWDFIT images for the energy range from 15–30 keV; the overall size of the source clearly increases steadily with energy. In [206] this trend was explained in terms of the increased penetration of higher-energy nonthermal electrons accelerated in the coronal part of the magnetic loop(s) involved in the flare. However, [52] showed that the increase in the overall length parameter with energy instead derives primarily from consideration of the two chromospheric

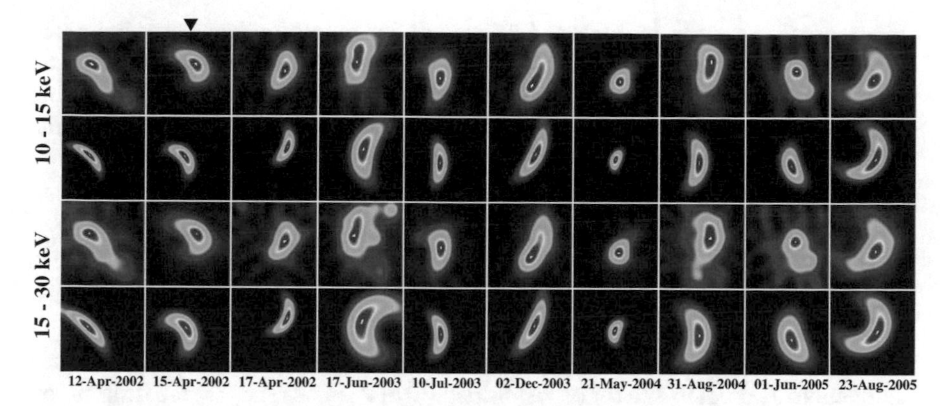

Fig. 6.2 CLEAN and visibility-based forward-fit images for the events studied in [206]. The first and third rows show the CLEAN images in the energy ranges 10–15 keV and 15–30 keV, respectively, while the second and fourth rows show the maps obtained by forward-fitting the observed visibilities to those corresponding to the curved elliptical Gaussian form (Eq. 6.1) in the same energy ranges. After [206], © AAS. Reproduced with permission.

footpoints, not clearly resolved in the CLEAN images and therefore not included in the VIS_FWDFIT analysis in [206]. Since the widely-separated footpoints have a harder spectrum than the coronal emission between them, estimation of an overall length parameter under the assumption that the source is a single curved elliptical source, results in an increase in that length parameter with energy. The implications of these results for solar flare physics are discussed in Chap. 7.

6.4 Bayesian Optimization

Bayesian optimization (or Bayesian parametric imaging) can be interpreted as a generalization of the forward-fit approach. Indeed, as in forward-fit, Bayesian imaging aims at establishing a relationship between a number of discrete sources modeled according to specific parametric shapes, and the observed visibilities. However, in contrast to forward-fit, this approach allows generalization in two ways.

- The number of sources is considered as one of the unknowns of the problem. The Bayesian algorithm thus quantitatively assesses the probabilities of different numbers of discrete sources within an image;
- The method provides samples of the probability distribution functions associated with the source parameters. This allows a more complete assessment to be made of the uncertainties on each parameter through analysis of higher moments of the probability distributions.

An effective way to implement Bayesian optimization for reconstruction problems is represented by SMC samples [169]. This technique realizes an automatic sampling of the posterior distribution associated with the solution of the imaging problem. The theoretical framework for SMC is given by the Bayes theorem, which, in the visibility framework[4] can be written as (see Sect. 4.3)

$$p(I|V) = \frac{p(V|I)\, p(I)}{p(V)} . \tag{6.2}$$

The different components in this equation are as follows:

- the likelihood distribution, $p(V|I)$, denotes the probability that the source function, I, predicts the set of visibilities V. $p(V|I)$ depends on the distribution of the noise affecting the visibilities and on the visibility formation model;
- the prior distribution $p(I)$ encodes all the information a priori known on the solution;
- $p(V)$ is a normalization constant; and
- the posterior distribution $p(I|V)$ is the solution of the parametric imaging problem. From it, one can compute the *maximum a posteriori argument*

$$I_{map} = \arg\max_I p(I|V) , \tag{6.3}$$

the *conditional mean*

$$I_{CM} = \int I\, p(I|V)\, dI , \tag{6.4}$$

and similarly all higher order (conditional) moments.

In parametric imaging, the image I to reconstruct can be modeled as

$$I = I(N, T_{1:N}, \Theta_{1:N}),$$

where N is the number of sources in the image,

$$T_{1:N} = (T_1, \ldots, T_N)$$

represents the source types (e.g., circles, ellipses, loop-like curved ellipses) and

$$\Theta_{1:N} = (\theta_1, \ldots, \theta_N)$$

contains the parameters characterizing each source. Figure 6.3 shows typical examples of the source shapes that can be used.

[4] An analogous approach can be realized when the measurements are the recorded counts.

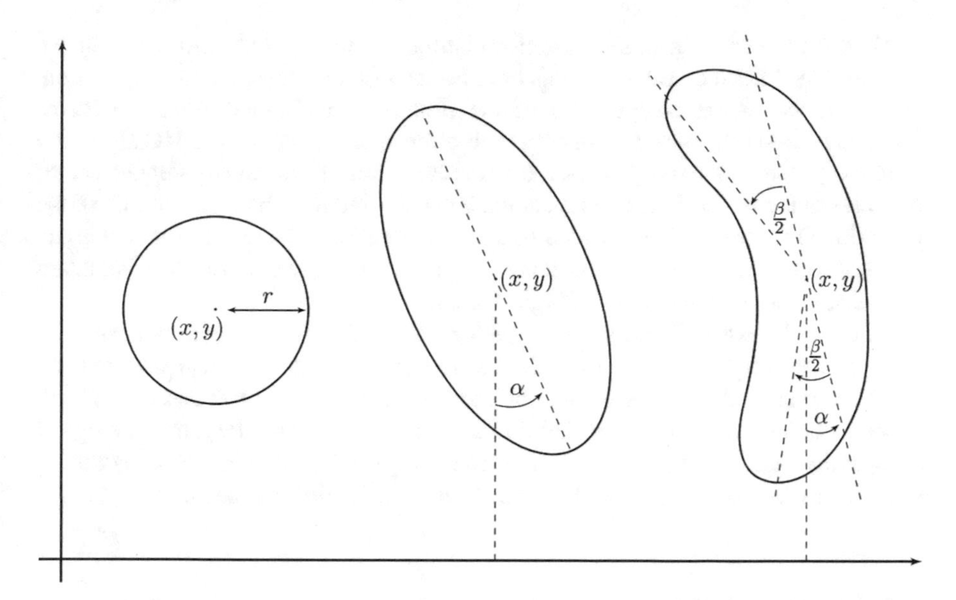

Fig. 6.3 SMC source shapes and their associated parameters. *left*: circle; *middle*: ellipse; *right*: curved ellipse. After [169], © AAS. Reproduced with permission.

A possible choice[5] for the prior distribution $p(I)$ is a factorized probability distribution function in which one of the factors is a Poisson distribution for N, and the other factors are uniform distributions for the source types and uniform distributions for the source parameters.

The likelihood function $p(V|I)$ is designed by relying on the information available on the model of data formation and on the statistical distribution of the noise affecting the observations. In the case of visibility-based imaging, data formation is mimicked by means of the Fourier transform, while a standard approximation is that the noise on the visibilities is Gaussian additive. It is therefore reasonable to assume that

$$p(V|I) \propto \exp\left(-\sum_i \frac{|V - F(N, T_{1:N}, \Theta_{1:N})|_i^2}{2\sigma_i^2}\right), \tag{6.5}$$

where $F(N, T_{1:N}, \Theta_{1:N})$ is the function of the imaging parameters $(N, T_{1:N}, \Theta_{1:N})$ that is obtained by inserting $I(N, T_{1:N}, \Theta_{1:N})$ in the visibility formation model, and the standard deviations σ_i on the visibilities are obtained from the Poisson statistics of the measured count rates.

[5] Of course, other choices are possible, depending on the a priori information one has at one's disposal.

Bayes' theorem relates the posterior distribution to the analytical forms of the prior and likelihood distributions that have been chosen for this imaging application. However, the actual computation of the posterior distribution from the Bayes formula is difficult, since it involves a high-dimensional space. The SMC algorithm iteratively computes the posterior distribution, thus producing a sample set of particles in the multi-dimensional parameter space and distributed according to the posterior. Once the SMC scheme terminates its iterations, the resulting samples can be used to compute the solution image, all image parameters with their standard deviations, and even moments of higher order.

In Sect. 7.1 we will illustrate the performance of SMC through application to the visibility bags observed by *RHESSI* during the C7.5 flare that occurred on 2002 February 20, with peak time at around 11:06 UT. As discussed in [169], several studies had pointed out that this event is characterized by a morphological complexity that evolves with both time and energy, with non-simultaneous flaring of different foot-point sources, thus making it ideal for this approach.

6.5 MEM_NJIT and MEM_GE

Maximum Entropy is an image reconstruction method conceived to image a sparsely occupied field of view and is therefore particularly appropriate for achieving super-resolution effects. Although widely used in ordinary image deconvolution (see Sect. 4.1.1), MEM has also been formulated in radio astronomy for the analysis of observations in the spatial frequency domain, and two IDL codes (MEM_NJIT and MEM_GE) have been implemented in the SSW tree for image reconstruction from solar X-ray Fourier-based data. The mathematical basis of both algorithms is the constrained maximization of the entropy function

$$H = -\sum_{j=1}^{M} I_j \ln I_j - I_j \, , \tag{6.6}$$

where I_j is the signal content of pixel j. This maximization is performed subject to three constraints. The first constraint simply requires that all components of I be non-negative, i.e.

$$I_j \geq 0 \qquad \forall \, j \, . \tag{6.7}$$

The second constraint requires that the total flux F' in the recovered image match the measured overall flux, as determined by the detector count rates:

$$F - F' \equiv \sum_{j=1}^{M} I_j - F' = 0 \, . \tag{6.8}$$

The third constraint is related to the χ^2 function corresponding to the differences between the N_v observed visibilities and those corresponding to the recovered image:

$$\chi^2 = \sum_{i=1}^{N_v} \frac{| \tilde{V}_i - V_i |^2}{\sigma_i^2} - N_v = 0 , \tag{6.9}$$

where $\{\tilde{V}_i\}_{i=1}^{N_v}$ is the set of visibilities predicted by the source signal I and σ_i is the vector of uncertainties in the measured visibilities V_i.

The algorithm, as implemented in the MEM_NJIT IDL code [22], addresses the constrained maximum problem

$$\arg \max_{I \geq 0} \{H - \alpha F - \beta \chi^2\} , \tag{6.10}$$

where α and β are the Lagrange multipliers associated with the constraints (6.8) and (6.9). These two parameters are not fixed *a priori* but are instead updated during the maximization process. The main drawback of this approach (see [131]) is that the maximization problem (6.10) involves a functional which is not convex and therefore numerical schemes for its maximization may lead to unstable solutions. Further, the functional is characterized by two Lagrange multipliers, and the updating process for these quantities can be non-optimal. Therefore, in some cases MEM_NJIT can produce nonphysical reconstructions.

The MEM_GE algorithm [131] addresses both these drawbacks by providing a different formulation of the maximum entropy optimization problem. The maximization problem (6.10) is replaced with the minimization problem

$$\arg \min_{I \geq 0} \{\chi^2 - \lambda H\} , \tag{6.11}$$

under the flux constraint (6.8), where λ is a regularization parameter. The main advantage of this approach is that the optimization problem is now convex and therefore can be addressed by relying on several numerical methods whose convergence properties are well-established. Further, the functional has just one Lagrange multiplier, so that the regularization step is less computationally demanding.

In [131], an iterative scheme is adopted whereby at each iteration a gradient step minimizes χ^2 and a proximal step [43] maximizes the entropy H subject to both the positivity constraint (Eq. (6.7)) and the total flux constraint (Eq. (6.8)). After this second step, the algorithm is accelerated by computing a linear combination with the approximation corresponding to the previous iteration. A monotonicity check [18] is also performed at this stage.

Two technical issues are concerned with the implementation of MEM_GE. On the one hand, the regularization parameter λ can be determined by applying one of the optimization methods developed within the framework of regularization theory [67]. On the other hand, an estimate of the flux F' can be realized a priori, for example, by exploiting the information contained in the measured visibilities or by solving

$$\arg\min_{\mathbf{x}\geq 0} \chi^2 , \tag{6.12}$$

and summing up the pixel contents of the solution.

Figures 6.4 and 6.5 show three cases when MEM_GE performs better than MEM_NJIT. Although the fitting is similar in both cases, MEM_NJIT produces multiple unrealistic point sources, while MEM_GE reconstructions are more consistent with those produced by other image reconstruction methods.

6.6 uv_smooth

The uv_smooth method [133] was inspired by the gridding methods developed within the framework of radio interferometry in which a non-uniformly sampled (u, v) plane is transformed into a uniformly sampled one. The method used here involves augmentation of the measured visibilities with a large number of Fourier components that are "manufactured" by smoothly filling in "gaps" between the measured points in the (u, v)-plane shown in Fig. 6.1 for *RHESSI* and *STIX*. Specifically, the method uses the following two-step iterative process:

1. The observed visibilities $V(u, v)$ are interpolated using two-dimensional spline interpolation to generate a smooth continuum of visibilities within the circle in the (u, v) plane spanned by the available data (Fig. 6.1);
2. To reduce the ringing effects that result from an unconstrained Fourier transform inversion procedure and that introduce undesirable super resolution effects [151], image positivity is imposed using a Fast Fourier Transform (FFT)-based iterative method.

To illustrate how uv_smooth works, we consider the simple but plausible example of a solar hard X-ray source in the form of a circular two-dimensional Gaussian with standard deviation equal to $4''$ and centered[6] at (450, 445) arcsec on the Sun. The Fourier transform of this Gaussian function is then determined at the *RHESSI*-sampled (u, v) points (Fig. 6.1) and an appropriate level of Poisson noise added to each visibility value to generate a set of simulated "measured" visibilities.

[6] It should be noted that the phase of the visibilities is determined by the distance of the (arbitrarily chosen) phase center (x_o, y_o) from the center (x_1, y_1) of the Gaussian. Because of the factors multiplying the Fourier-space coordinates u and v in this expression, a large distance between the phase center and the image centroid produces a visibility function $V(u, v)$ that varies very rapidly, making construction of a useful set of discrete "observed" visibilities very challenging, if not impossible. Success of the method therefore depends on choosing the phase center at a point relatively close to the Gaussian centroid of the image.

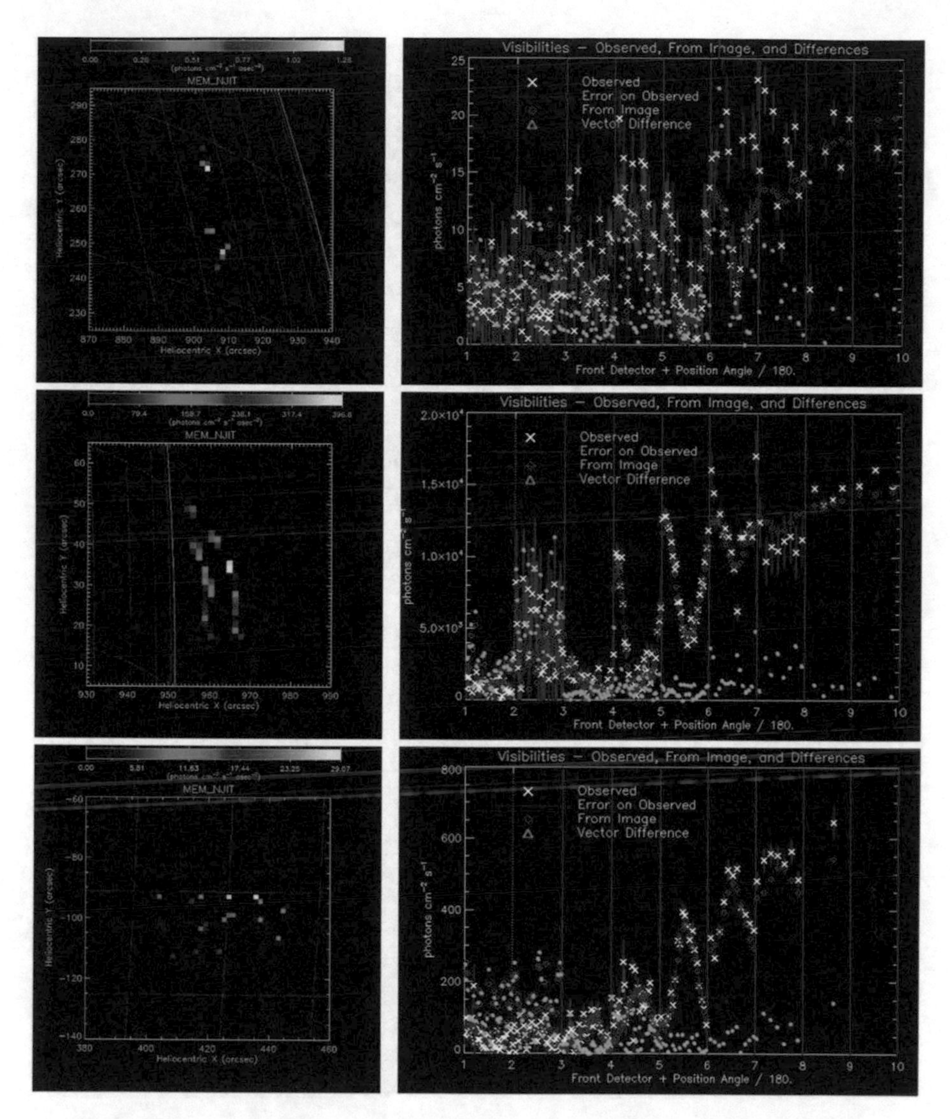

Fig. 6.4 *Left column*: Images made using MEM_NJIT of three flares observed with *RHESSI*: *Right column*: comparison between predicted and measured visibilities for these events. The three events are: the 2002 February 20 event, time interval 11:06:05–11:07:42 UT, energy range 25–50 keV, detectors 3 through 9 (first row); the 2002 May 1 event, time interval 19:21:29–19:22:29 UT, energy range 3–6 keV, detectors 3 through 9 (second row); the 2007 December 13 event, time interval 22:11:33–22:12:56 UT; energy range 6–12 keV; detectors 3 through 9 (third row). Note for all events the breakup into nonphysical compact sources, as the method attempts to obtain a better fit to the noisy visibility amplitudes. After [131], © AAS. Reproduced with permission.

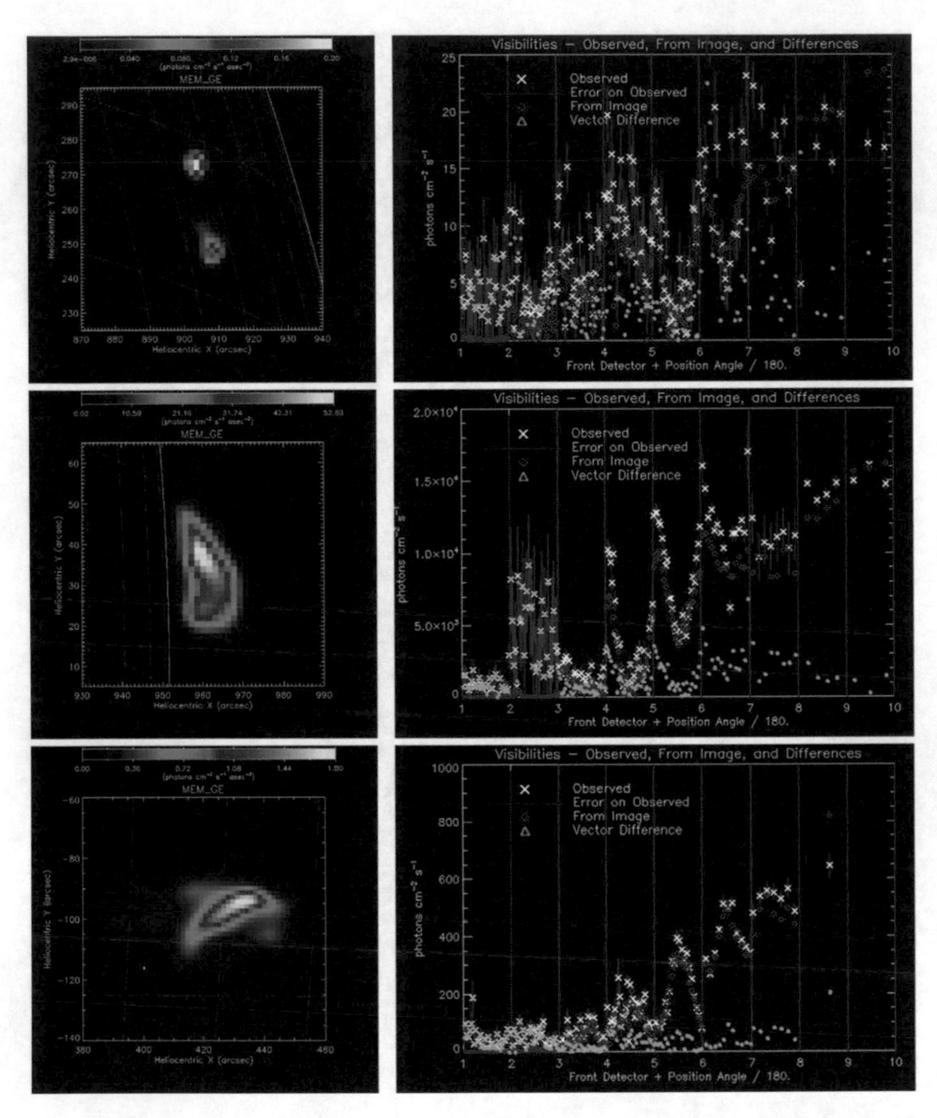

Fig. 6.5 Same as Fig. 6.4 for images of the same flares made with MEM_GE. Note the more physical source shapes for all events. After [131], © AAS. Reproduced with permission.

In the tests of the uv_smooth method [133], the source width is somewhat greater than the angular resolution of the finest *RHESSI* collimator, but only visibilities corresponding to *RHESSI* detectors #3 to #9 were used, corresponding to spatial resolutions of $\alpha_3 \sim 7''$ to $\alpha_9 \sim 180''$, or, equivalently, visibility magnitudes $\sqrt{u^2+v^2}$ from $1/2\alpha_9 \simeq 0.0027$ arcsec^{-1} to $1/2\alpha_3 \simeq 0.073$ arcsec^{-1}.

Figure 6.6 shows how the interpolation step of uv_smooth works with different interpolation schemes [133]. The images in rows (a) and (b) use a back-projection

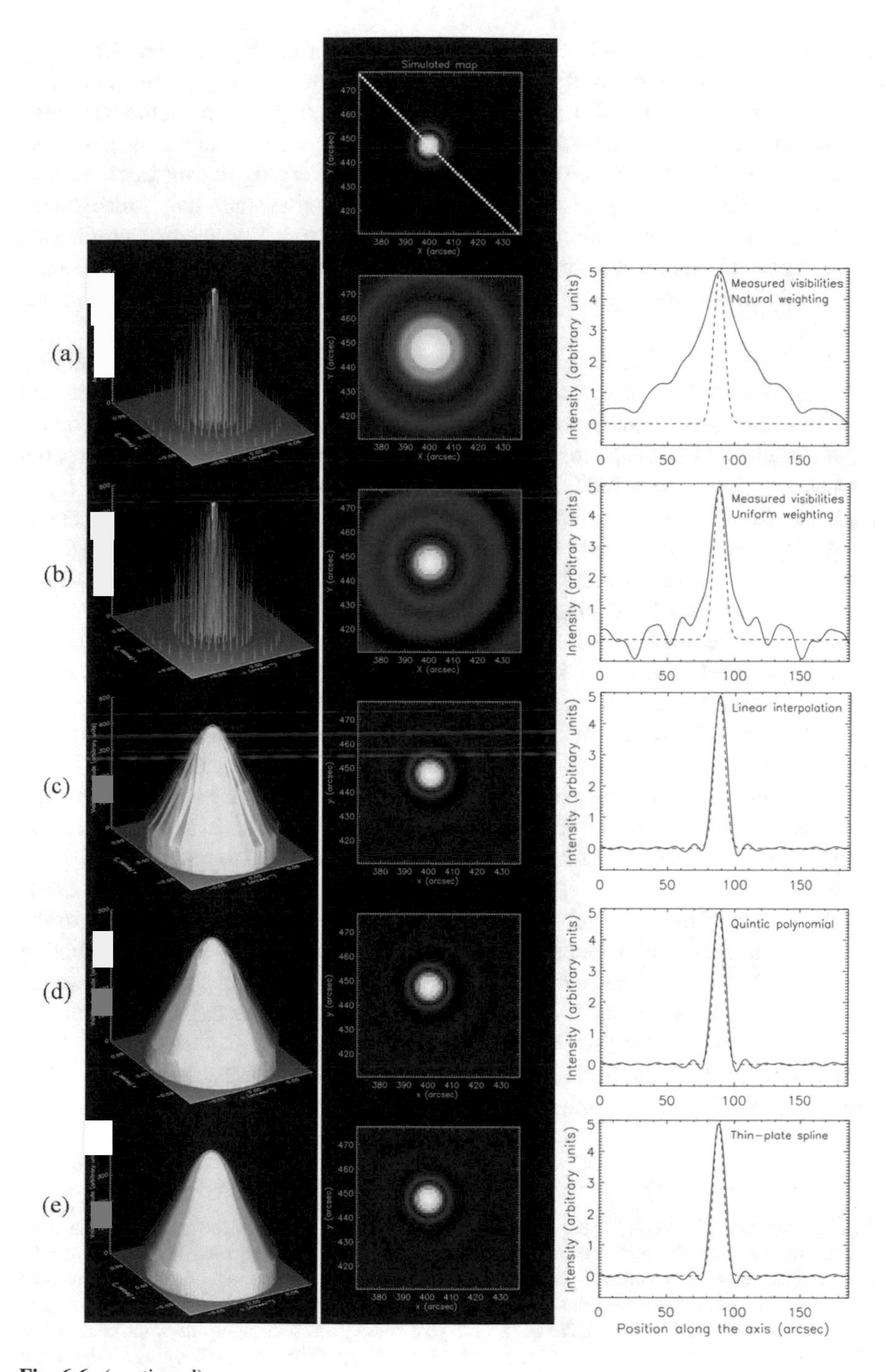

Fig. 6.6 (continued)

algorithm with natural and uniform weightings, respectively, that employs only the discrete set of "measured" visibilities. Rows (c) through (e) are the smoothed visibility surfaces, generated using three different interpolation methods: linear, quintic polynomial, and thin-plate spline [133], respectively. Radial cuts across the source structure (right column of Fig. 6.6) reveal the very significant improvement in image quality that is obtained when interpolated, rather than only "measured," visibilities are used. In particular, the plate-spline interpolation method produces a large degree of overall smoothness in the visibility surface, with the added feature that the smoothed $V(u, v)$ surface passes through the "measured" visibility points *exactly*.

Despite the considerable improvement produced by these smoothed visibility surfaces, it must be noted that the smoothed surfaces still cover only a limited portion of the (u, v) plane, namely the annulus between the innermost and outermost circular loci corresponding to *RHESSI* detectors #3 through #9, i.e., spatial resolutions of $\sim 7''$ to $\sim 180''$. In particular, limited by the resolution of the finest grid pair (the pair in front of detector #3) used in the reconstruction, there is no (high-frequency) visibility information outside the disk of radius $\rho = R_3 \simeq 0.073$ arcsec^{-1}. Without this high-frequency information, it is impossible to fully realize the true form of the original Gaussian image.

This high-frequency limit results both in an overestimation of the size of the reconstructed image and in the appearance of residual "ringing" oscillations that result from truncation of Fourier-space information. To reduce the "ringing" oscillations, it is necessary to add a second extrapolation step, in which a positivity constraint is imposed using the projected Landweber method [152], an FFT-based iterative method. Enforcing positivity at each point in this manner does have the additional undesirable effect of slightly increasing the total flux in the image, and to counter this effect, the total flux in the image is renormalized to its measured value at each step.

The results of this two-step method are illustrated in Fig. 6.7. Compared to the interpolation-only results (center column of Fig. 6.6), there is a noticeable reduction of the ringing effects. The method produces the desired result: a visibility surface (top panel of column (b) of Fig. 6.7) that is positive everywhere but which gradually tapers to zero outside the limiting radius $\rho = R_3$ of the data.

The top panel of column (c) of Fig. 6.7 shows a radial cut through the visibility surface. The originally sampled visibilities are shown as open squares, while the

Fig. 6.6 (continued) A comparison of visibility interpolation methods for a circular Gaussian source. *Topmost panel:* an image of the original source. The next five rows show (left panels) the visibility amplitudes, (middle panels) the corresponding recovered images, and (right panels) radial intensity profile for the imaged source (solid line) compared with that corresponding to the original source (dashed line). *Rows* (**a**) and (**b**): back-projection algorithm with natural and uniform weighting, respectively; *Row* (**c**): linear interpolation method; *Row* (**d**): quintic polynomial interpolation method; *Row* (**e**): thin-plate spline algorithm. After [133], © AAS. Reproduced with permission.

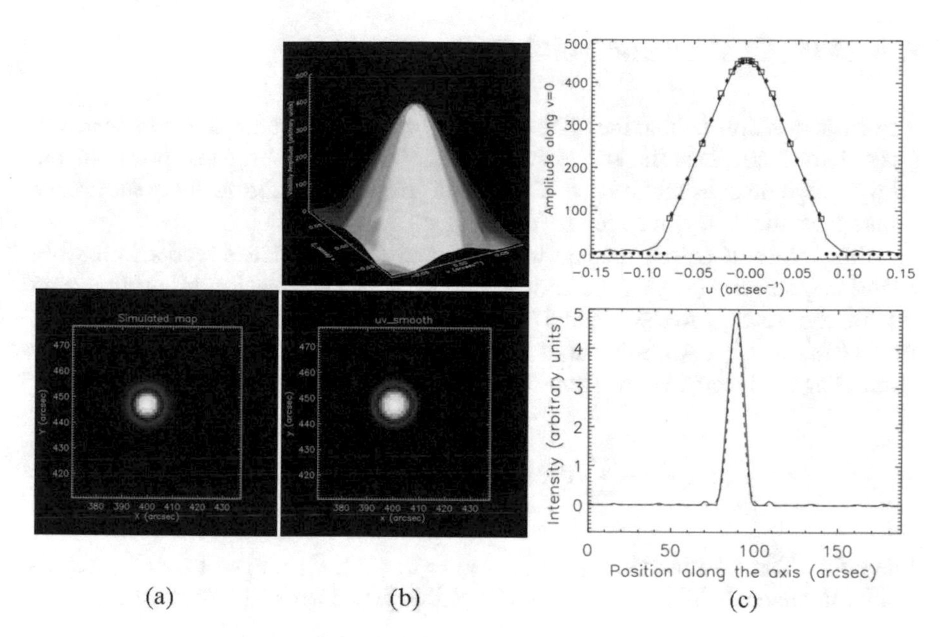

Fig. 6.7 Imposing the positivity constraint. *Column* (**a**): the original synthetic image. *Column* (**b**): the visibility surface (upper panel) and image (lower panel) corresponding to the final step of the iterative method. *Column* (**c**), *upper panel:* The solid line shows a radial cut through the origin of the visibility surface. The originally sampled (measured) visibilities are shown as open squares, while the interpolated visibilities are shown as points (with $\pm 1\sigma$ error bars). The solid dots represent a zero-padding outside the range of the sampled data. *Column* (**c**), *lower panel:* The solid line shows the same radial cut through the image as in Fig. 6.6, compared to the same cut (dashed line) through the original source. After [133], © AAS. Reproduced with permission.

interpolated visibilities are shown as points (with $\pm 1\sigma$ error bars). Despite the fact that imposing positivity results in visibilities that are no longer *exactly* equal to the measured values at the sampled (u, v) points, the visibility surface is still an acceptable fit to the data. The lower panel of column (c) of Fig. 6.7 is a radial cut across the corresponding spatial image; compared to the cuts shown in Fig. 6.6, we see that the ringing is strongly suppressed. The method reproduces well the location and intensity of the peak and, although it overestimates the FWHM source width by some 15%, this is an inescapable manifestation of the instrumental point-spread function, common to all image reconstruction methods.

Finally, we point out that more recent versions of the uv_smooth approach exploit feature-augmentation processes inspired by artificial intelligence that notably improve the interpolation step relying on prior information on the flaring source [149].

6.7 VIS_CLEAN and Multi-Scale CLEAN

The basic principle behind the CLEAN algorithm has been summarized in Sect. 5.2. Essentially it involves the successive identification of the brightest pixels of the "dirty" map and construction of a "CLEANed" map by transferring these successive points from the "dirty" map to the "CLEANed" map.

Although the CLEAN method was designed for, and indeed works well with, actual maps in the (x, y) domain, it is important to note that, with Fourier-based instruments such as *RHESSI* and *STIX*, the "native" data are N_V measured visibilities $\{V(u_l, v_l)\}_{l=1}^{N_V}$. An estimate of the dirty map can be obtained by numerically computing the inverse Fourier transform of the set of sampled visibilities, i.e.,

$$g(x, y) = \sum_{l=1}^{N_V} V(u_l, v_l) \exp\left[-2\pi i\, (xu_l + yv_l)\right] . \tag{6.13}$$

Since for a Dirac δ-function source $I(x, y) = \delta(x, y)$, $V(u, v) = 1$ for all sampled spatial frequencies (u, v) (Eq. (3.5)). Hence the dirty beam can be computed as

$$K(x, y) = \sum_{l=1}^{N_V} \exp\left[-2\pi i\, (xu_l + yv_l)\right] . \tag{6.14}$$

It follows that *the dirty beam is the discretized inverse Fourier transform of the characteristic function of the sampled visibilities* (i.e., the transfer function of the imaging system). An idealized version of the PSF, the CLEAN beam, may be obtained by fitting the $K(x, y)$ calculated using Eq. (6.14) with a two-dimensional Gaussian function. With these specific definitions for the dirty map, the dirty beam, and the CLEAN beam, the general mathematical setup for the CLEAN algorithm described in Sect. 5.2 can be used. This is the basis for the VIS_CLEAN formulation.

Both the *RHESSI* and *STIX* telescopes are characterized by a PSF that can be written, under some reasonable assumptions, as the sum of a finite number of PSF components, each one filtering a specific portion of the (u, v) plane. For example, *RHESSI* samples visibilities over nine circles in the (u, v) plane with radii decreasing from $R_1 \simeq 0.22$ arcsec^{-1} to $R_9 \simeq 0.0027$ arcsec^{-1}, according to a geometric progression with ratio $1/\sqrt{3}$ (Fig. 6.1). On the other hand, as discussed in Sect. 3.7, *STIX* samples the (u, v) plane in 60 points placed over six spirals (Fig. 3.7) in such a way that these points lie on 10 circles with radii decreasing from $\simeq$0.07 arcsec^{-1} to $\simeq$0.0028 arcsec^{-1}, according to a geometric progression with ratio $\simeq$0.7. Therefore we can define a number N of experimental PSF components, i.e., a number of dirty beams $\{K_j(x, y)\}_{j=1}^{N}$, such that

$$K_j(x, y) = \sum_{l=1}^{N_j} \exp\left[-2\pi i\, (xu_l^j + yv_l^j)\right] \quad j = 1, \ldots, N , \tag{6.15}$$

where N_j and $\{u_l^j, v_l^j\}_{l=1}^{N_j}$ are the number, and the set of visibilities belonging to, the jth subset of sampled circles in the (u, v) plane, respectively. For example, for *RHESSI* we have up to $N = 9$ such "composite" dirty beams, each of which is a combination of the dirty beams for the (u, v) points on each of the nine circles sampled by *RHESSI*. Similarly, for *STIX* we have up to $N = 10$ "composite" dirty beams. Further, since $\sum_{j=1}^{N} N_j = N_V$, then

$$K(x, y) = \sum_{j=1}^{N} K_j(x, y) . \tag{6.16}$$

Using this approach, we can define a set of N dirty maps

$$g_j(x, y) = \sum_{l=1}^{N_j} V(u_l^j, v_l^j) \exp[-2\pi i (x u_l^j + y v_l^j)] \quad j = 1, \ldots, N \tag{6.17}$$

such that

$$g(x, y) = \sum_{j=1}^{N} g_j(x, y) . \tag{6.18}$$

The corresponding model for the source image is the sum of N basis functions $\{m_i\}_{i=1}^{N}$, each characterized by a specific scale. The source image $I(x, y)$ can be modeled as the superposition of these basis functions (plus background), i.e.,

$$I(x, y) = \sum_{i=1}^{N} \sum_{q_i=1}^{Q_i} I_{q_i} m_i(x - x_{q_i}, y - y_{q_i}) + B(x, y) , \tag{6.19}$$

which means that at scale i, for $i = 1, \ldots, N$, there are Q_i sources, each one placed at (x_{q_i}, y_{q_i}) and with peak intensity I_{q_i} for $q_i = 1, \ldots, Q_i$. Inserting Eq. (6.19) into the model Equation (5.2) leads to

$$g(x, y) = \sum_{i=1}^{N} \sum_{q_i=1}^{Q_i} I_{q_i} (m_i * K) (x - x_{q_i}, y - y_{q_i}) + (K * B)(x, y) . \tag{6.20}$$

By exploiting Eqs. (6.16) and (6.18), we obtain

$$\begin{aligned} \sum_{j}^{N} g_j(x, y) &= \sum_{j=1}^{N} \sum_{i=1}^{N} \sum_{q_i=1}^{Q_i} I_{q_i} (m_i * K_j)(x - x_{q_i}, y - y_{q_i}) + \sum_{j=1}^{N} (K_j * B) (x, y) \\ &= \sum_{j=1}^{N} \left(\sum_{i=1}^{N} \sum_{q_i=1}^{Q_i} I_{q_i} (m_i * K_j) (x - x_{q_i}, y - y_{q_i}) + (K_j * B) (x, y) \right) . \end{aligned} \tag{6.21}$$

This equation inspires a multi-scale version of the CLEAN algorithm in which each dirty map $g_j(x, y)$ is identified as

$$g_j(x, y) = \sum_{i=1}^{N} \sum_{q_i=1}^{Q_i} I_{q_i} (m_i * K_j) (x - x_{q_i}, y - y_{q_i}) + (K_j * B) (x, y) \qquad (6.22)$$

and cleaned accordingly. It may be seen that the key idea is based on the interplay between the basis functions and the PSF components. Equation (6.22) shows that each dirty map contains cross-convolution products between the corresponding PSF component and all the basis functions. Computing these cross-convolution products in the (u, v) plane, one immediately notes that $m_i * K_j$ with $i > j$ are negligible. The strategy is therefore to identify small scale sources first and transfer them from the set of dirty maps to the cleaned map.

In the case of *RHESSI*, the global PSF can be written as the sum of up to nine components, each one corresponding to a specific sub-collimator. Therefore, this global PSF can be decomposed, for instance, as the sum of three PSF components, the first one associated with the two sub-collimators with finest angular resolution (sub-collimator #s 1 and 2), the second one associated to the three sub-collimators with intermediate resolution (sub-collimator #s 3, 4, and 5), and the third one associated to the four sub-collimators with the coarsest resolution (sub-collimator #s 6, 7, 8, and 9). Correspondingly, one can construct three basis functions, which are three two-dimensional Gaussian functions with standard deviations determined by the three PSF components. These computational tools allow one to illustrate the potential of multi-scale CLEAN in the case of a simulated test relying on synthetic *RHESSI* visibilities. The central top panel of Fig. 6.8 is a source configuration inspired by the event observed by *RHESSI* on 2003 December 2, in the time interval 22:54:00–22:58:00 UT, in the energy range between 16 and 18 keV. Three visibility sets have been generated, corresponding to three different levels of progressively improved statistical uncertainty (an average of 1000, 10,000 and 100,000 counts per detector, respectively). Finally, multi-scale CLEAN, based on the three different resolution scales, is applied to obtain the reconstructions in the panels of the third row of Fig. 6.8 for each of the three levels of statistical uncertainty. The use of multiple scales clearly improves the spatial resolution when compared to the reconstructions provided by standard CLEAN (middle row panels of the same figure).

6.8 Compressed Sensing: VIS_CS and VIS_WV

The basic idea of compressed sensing in image processing is that an image can be effectively reconstructed from a relatively small number of indirect measurements if the image can be represented as the linear superposition of a similarly small number of basis functions. Denoting as BI such a linear transformation of the

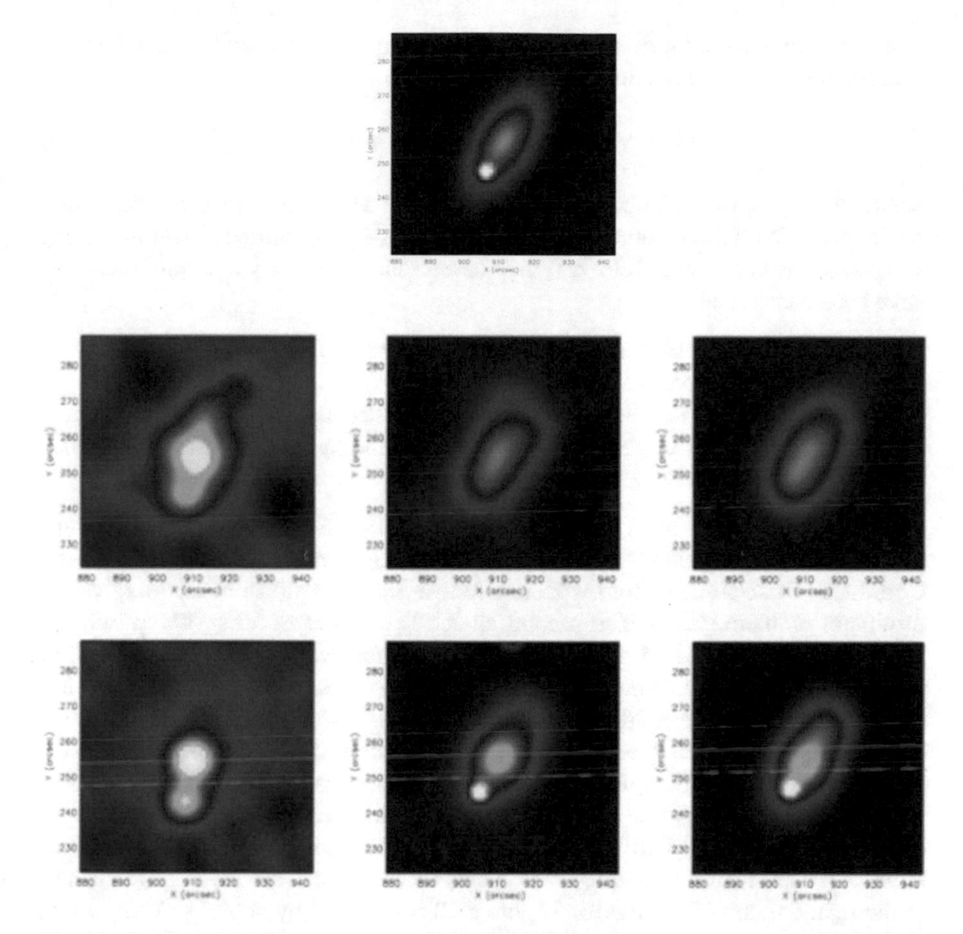

Fig. 6.8 Multi-scale CLEAN applied to synthetic visibilities corresponding to the configuration in the top central panel; middle row: reconstructions provided by standard CLEAN in the case of low, medium and high statistics; bottom row: reconstruction provided by multi-scale CLEAN for the same levels of statistics.

image I to reconstruct, compressed sensing is, in principle, realized by solving the minimization problem

$$\|V - \chi \cdot \mathcal{F}I\|_2^2 + \lambda \|BI\|_0 = \text{minimum}\,, \tag{6.23}$$

where V is the vector of the measured visibilities, $\mathcal{F}$ is the Fourier transform, χ is the characteristic function of the set of points of the (u, v) plane sampled by either *RHESSI* or *STIX*, $\| \cdot \|_0$ denotes the $L0$-norm that counts the total number of non-zero elements in the vector BI, and the regularization parameter λ tunes the trade-off between sparsity of the data and the accuracy of the fit. It is, however, well-established that (6.23) is a computationally inaccessible combinatorial problem.

Therefore, compressed sensing is, in practice, approximated by addressing the more computationally-feasible minimization problem

$$\|V - \chi \cdot \mathcal{F} I\|_2^2 + \lambda \|BI\|_1 = \text{minimum} \ , \tag{6.24}$$

where $\| \cdot \|_1$ is the $L1$-norm that sums up the absolute values of the vector components. With this change, the compressed sensing algorithm reduces to making an appropriate choice for B. In visibility-based solar hard X-ray imaging, two cases have been considered so far:

- A sample of up to 10^6 basis functions randomly sampled from the infinite set of Gaussian distributions [69]; and
- A Finite Isotropic Wavelet transformation, constructed as the two-dimensional isotropic extension in the Fourier domain of the symmetric one-dimensional Meyer mother function [57].

The first choice has been implemented in the VIS_CS code, which is one of the methods utilized to generate the *RHESSI* flare image archive.[7] VIS_CS utilizes a Coordinate Descent algorithm with the Active Set heuristic in order to solve the minimum problem (6.24). The second choice is the core of VIS_WV, which has been extensively validated against both *RHESSI* observations and synthetic *STIX* data. The optimization of problem (6.24) in VIS_WV is realized by using the fast iterative shrinkage-thresholding algorithm (FISTA). Both codes are available in the SSW tree and are accessible from both the IDL command line and the HESSI GUI.

Examples of how compressed sensing works in this setting are illustrated in Figs. 6.9 and 6.10. Figure 6.9 compares VIS_CS and CLEAN in the case of synthetic simulations of different source shapes and count statistics (these three simulations are inspired by three real events whose shapes are given in the panels at the right column of the figure). Figure 6.10 is obtained by applying VIS_WV to visibility bags associated with the 2002 July 23 event, for several energy channels.

6.9 Electron Flux Maps

As discussed in Chap. 1, the fundamental science goal behind hard X-ray imaging spectroscopy observations is to obtain information on the flux spectrum of the accelerated electrons throughout the flaring volume. In the traditional approach to imaging spectroscopy, the first step toward this goal is to use the techniques discussed in Chap. 5 and/or in the other sections of this chapter to obtain spatial images at different count energies. In principle, these images can then be "stacked" to produce count spectra at selected regions $(x \pm \Delta x, y \pm \Delta y)$ within the image. Application of a regularized count $\rightarrow$ electron flux inversion, using both the

[7] https://hesperia.gsfc.nasa.gov/rhessi_extras/flare_images/hsi_flare_image_archive.html.

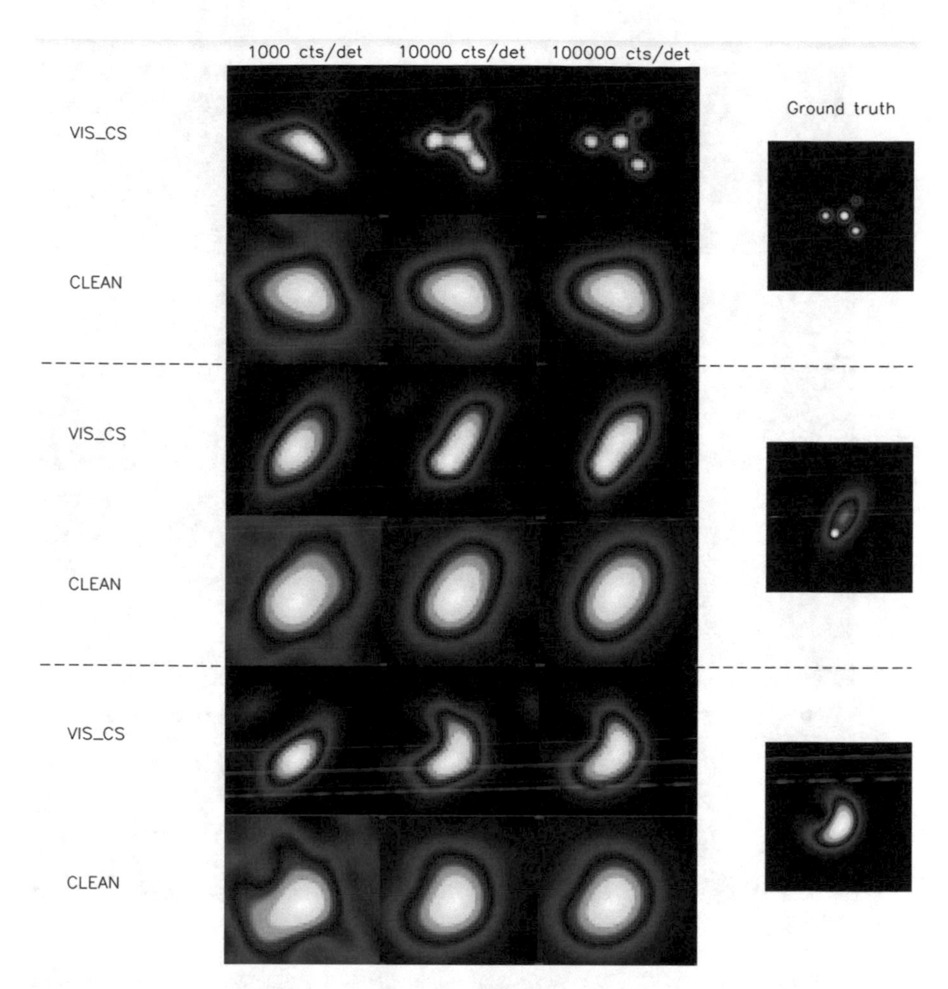

Fig. 6.9 Performances of VIS_CS: starting from the three realistic flare configurations in the last column (ground truth) three visibility sets have been generated corresponding to three levels of statistic. The reconstructions provided by VIS_CS are compared to the ones provided by the default version of CLEAN. After [69],© AAS. Reproduced with permission.

spectrometer response matrix and the bremsstrahlung cross-section, is then used to obtain the mean source electron flux spectrum (Eq. (1.12)) for the region selected. With some assumptions about the thick-target or thin-target nature of the electron interactions in the solar atmosphere, the spectrum of the accelerated electrons can then be determined. In practice, thick- and thin-target routines (called thick2 and thin2, respectively) have been developed that allow the accelerated electron flux spectrum to be determined directly from the count spectrum in each sub-region.

However, such an approach is not entirely ideal. As we have seen, the count-based images produced typically contain a substantial amount of noise, and, although, as discussed above, various algorithms have been developed to produce

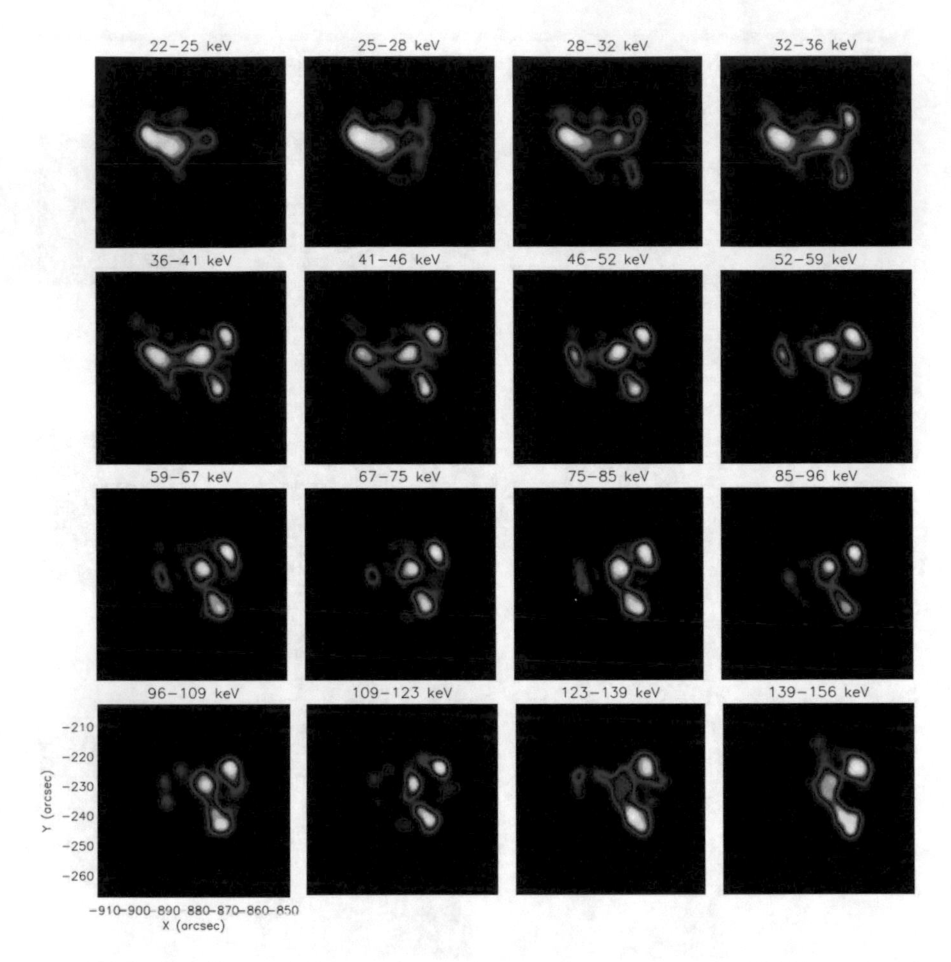

Fig. 6.10 Performance of VIS_WV when applied to *RHESSI* measured visibilities recorded during the 2002 July 23 event in several energy channels between 22 and 156 keV. After [57], © AAS. Reproduced with permission.

more useful images, an unavoidable level of data noise necessarily remains. Most importantly, each image is produced using a count set that is completely independent of those used to make images in adjacent energy bins, and so recovered images corresponding to adjacent count energy bins can exhibit substantial differences from each other. "Stacking" such noisy images leads to substantial bin-to-bin fluctuations in the count spectrum for a particular sub-region ($x \pm \Delta x$, $y \pm \Delta y$), and the ill-posedness of the spectral inversion problem (Sect. 4.1) means that spectral inversion of such a noisy count spectrum can be very problematic. Making this situation worse is that, as pointed out in Eq. (1.11) and subsequent discussion, *any* mean source electron spectrum $\langle n V F(E) \rangle$ necessarily produces a photon spectrum that is a monotonically decreasing function of photon energy ϵ. The noisy count spectra obtained by stacking of independent noisy images may not satisfy this fundamental property, so that the corresponding physically possible electron spectra may not even exist [154].

A resolution of this issue [157] follows from the essence of the manner in which *RHESSI* produces images, namely the conversion of temporal modulation of flux into spatial information in the Fourier (spatial frequency) domain. This means that photon (or, more accurately, count) spectra are most readily obtained not at each point in the direct spatial domain, but rather at each point in the *spatial frequency* domain. Hence the optimum way of obtaining spatial information on the spatial variation in the spectrum of accelerated electrons is to first invert the observed count spectra *obtained at points in the spatial frequency domain* using a regularized *spectral* inversion procedure [e.g., 150], thereby obtaining smoothed electron energy spectra for each observed spatial frequency point (u, v). Once such *electron flux visibility spectra* have been obtained, they can subsequently be (inverse) Fourier-transformed to yield "electron flux images," from which one can then obtain the electron flux at any sub-region of interest. Because of the regularized spectral inversion process used to obtain the electron flux visibility spectra, they necessarily vary smoothly with electron energy E. Hence, when the electron images are "stacked" to produce corresponding electron flux spectra at each location in the image, such spectra will also vary smoothly with energy. As we shall see in Sect. 7.4, this property allows us to derive substantial insight into the underlying physics of the accelerated electrons.

Essentially, then, we interchange the order of two steps in the data processing chain (see Fig. 6.11). From a formal mathematical point of view, this reversal of order is permitted because of the inherent linearity of both the spatial (Fourier transform) and spectral (regularized inversion) operations. Following [154], we now present a formal development of the necessary quantities and their interrelationships.

Define a Cartesian coordinate system (x, y, z) (in units of cm) such that (x, y) represents a location in the image plane and z represents distance along the line of sight. Let the local density of target particles within a source of line-of-sight depth ℓ be $n(x, y, z)$ (cm^{-3}) and let the differential electron flux spectrum (electrons cm^{-2} s^{-1} keV^{-1}) at the point (x, y, z) in the source be $F(x, y, z; E)$. We formally define the *mean source spectral flux image* $\overline{F}(x, y; E)$ (electrons cm^{-2} s^{-1} keV^{-1}) by

$$\overline{F}(x, y; E) = \frac{1}{N(x, y)} \int_{z=0}^{\ell(x,y)} n(x, y, z)\, F(x, y, z; E)\, dz\,, \qquad (6.25)$$

where the column density (cm^{-2}) at each point (x, y) in the image is given by $N(x, y) = \int_{z=0}^{\ell(x,y)} n(x, y, z)\, dz = \overline{n}(x, y)\, \ell(x, y)$. Since the source is optically thin along its entire extent, the relation between $\overline{F}(x, y; E)$ and the corresponding

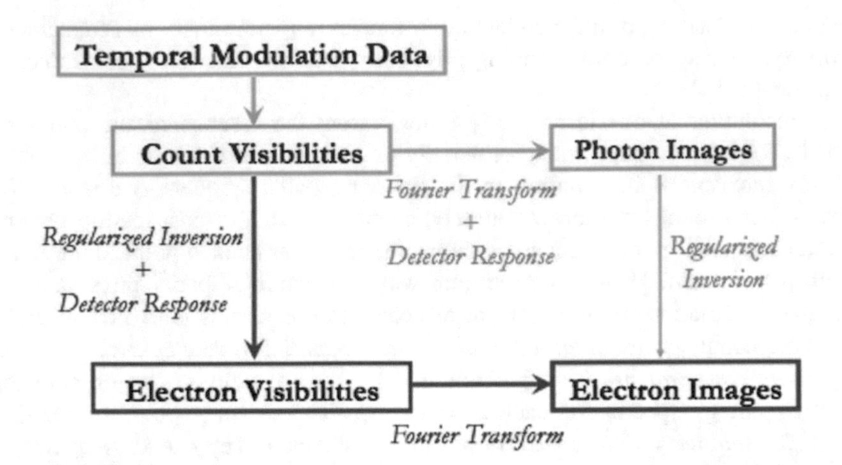

Fig. 6.11 The electron visibility concept. Proceeding from count-based visibilities to images of the electron flux involves two inversions, one spatial and one spectral. Traditionally, the spatial (Fourier) inversion is performed first, and then the count spectrum at a given location (formed by "stacking" count-based images at different count energies) is subjected to a regularized spectral inversion to yield the electron flux spectrum at that location. However, uncorrelated statistical uncertainties from image to image make the count spectrum rather noisy, and even nonphysical. It is therefore expedient to reverse the order of spectral and spatial inversion, first performing a spectral inversion on the count visibilities to yield *electron flux visibilities*, and then constructing electron flux images from these visibilities using standard visibility-based image reconstruction techniques. Stacking these images produces the electron flux spectrum at each sub-region in the image plane. This method has the added advantage that the electron visibilities are regularized and so vary smoothly with energy; accordingly, the electron flux maps (see Fig. 6.12 below) also vary smoothly with energy.

photon image $I(x, y; \epsilon)$ (photons s^{-1} keV^{-1} per cm^2 of flare area per cm^2 of detector area) is (cf. Eq. (1.11))

$$I(x, y; \epsilon) = \frac{1}{4\pi R^2} \int_{E=\epsilon}^{\infty} N(x, y)\, \overline{F}(x, y; E)\, Q(\epsilon, E)\, dE \;, \tag{6.26}$$

where R equals 1 AU and $Q(\epsilon, E)$ (cm^2 keV^{-1}) is the bremsstrahlung cross-section[8] for photon emission, differential in photon energy ϵ.

We next define the *differential count visibility* $V(u, v; q)$ (counts s^{-1} keV^{-1} per cm^2 of detector area) by

$$V(u, v; q) = \int_x \int_y J(x, y; q)\, e^{i(ux+vy)}\, dx\, dy \;. \tag{6.27}$$

[8] The form of the quantity $Q(\epsilon, E)$ depends on the emission process being considered. In principle, this could include a host of emission processes, such as gyrosynchrotron emission, inverse Compton emission, free-bound emission, and bremsstrahlung. We here take the form of $Q(\epsilon, E)$ as that corresponding to bremsstrahlung, and we use the isotropic form of the cross-section in [105].

Incorporating the photon → count detector response matrix, we may write

$$V(u, v; q)\, dq = \int_x \int_y \int_{\epsilon=q}^{\infty} D(q, \epsilon)\, I(x, y; \epsilon)\, e^{i(ux+vy)}\, d\epsilon\, dx\, dy\,, \tag{6.28}$$

where $D(q, \epsilon)$ (keV^{-1}) is the differential element of the detector response matrix corresponding to the generation of a count with energy in the energy range $[q, q + dq]$ from a photon in the energy range $[\epsilon, \epsilon + d\epsilon]$.

Combining Eqs. (6.26) and (6.28) gives the rather formidable expression

$$V(u, v; q)\, dq = \frac{1}{4\pi R^2} \times$$

$$\int_x \int_y \int_{\epsilon=q}^{\infty} \int_{E=\epsilon}^{\infty} [N(x, y)\, \overline{F}(x, y; E)]\, D(q, \epsilon)\, Q(\epsilon, E)\, e^{i(ux+vy)}\, dE\, d\epsilon\, dx\, dy\,, \tag{6.29}$$

which provides the formal relationship between $[N(x, y)\, \overline{F}(x, y; E)]$, the quantity of most direct physical interest, and the observed differential count visibilities $V(u, v; q)$.

We now introduce the *differential count cross-section* $K(q, E)$ (cm^2 keV^{-1}) through the relation

$$K(q, E)\, dq = \int_{\epsilon=q}^{\infty} D(q, \epsilon)\, Q(\epsilon, E)\, d\epsilon\,, \tag{6.30}$$

and the *differential electron flux visibility* (electrons s^{-1} keV^{-1} per cm^2 of detector area)

$$W(u, v; E) = \int_x \int_y N(x, y)\, \overline{F}(x, y; E)\, e^{i(ux+vy)}\, dx\, dy\,. \tag{6.31}$$

With these definitions, and reversing the order of integration with respect to ϵ and E, Eq. (6.29) can be written as the integral equation

$$V(u, v; q) = \frac{1}{4\pi R^2} \int_q^{\infty} W(u, v; E)\, K(q, E)\, dE\,. \tag{6.32}$$

Equation (6.32) is formally identical to Eq. (6.26) and so can be solved for the visibilities $W(u, v; E)$ from the observed count visibilities $V(u, v; q)$ through the use of well tested (e.g., [150, 153]) Tikhonov regularization techniques (Sect. 4.3). Briefly, Eq. (6.32) is first discretized in both count and electron energy spaces to yield, at each sampled point (u, v) in the spatial frequency domain, the *data visibility vector* $\mathbf{V}_{[u,v]}$ (the elements of which depend on count energy q) and the *source visibility vector* $\mathbf{W}_{[u,v]}$ (the elements of which depend on electron energy E). These are related through the matrix equation

$$\mathbf{V}_{[u,v]} = \mathbf{K} \cdot \mathbf{W}_{[u,v]} \,, \tag{6.33}$$

where $\mathbf{K}$ is the kernel matrix, the elements of which are formed from the values of $K(q, E)$ at the discretized count and electron energy points. Then the *zero-order regularization problem*

$$||\mathbf{V}_{[u,v]} - \mathbf{K} \cdot \mathbf{W}_{[u,v]}||^2 + \lambda_{[u,v]}||\mathbf{W}_{[u,v]}||^2 = \text{minimum} \tag{6.34}$$

is solved for $\mathbf{W}_{[u,v]}$ given the prescribed visibility vector $\mathbf{V}_{[u,v]}$ at each sampled point in (u, v) space, using an appropriate value (see below) of the smoothing parameter $\lambda_{[u,v]}$. This results in electron visibility spectra that are "smooth" in the sense that the large variations in $W(u, v; E)$ from energy bin to energy bin are suppressed. This technique therefore enhances spatial features (Fourier components) that persist over a wide range of energies, and suppresses (noise) features that exist only over a rather narrow range of energy bins. Once the differential electron visibility spectra have been determined, mean source electron flux images may be determined through inversion of Eq. (6.31), namely

$$N(x, y)\, \bar{F}(x, y; E) = \int_u \int_v W(u, v; E)\, e^{-i(ux+vy)}\, du\, dv \,. \tag{6.35}$$

It is important to realize that, since the spectral inversion in Eq. (6.32) is performed using a regularized inversion procedure [e.g., 154], the electron visibilities will, by construction, vary more smoothly with energy than the count-based visibilities used to construct them. As a result, the electron flux maps generated from the electron visibilities will not only exist (cf. discussion at the beginning of this Section), but will also vary more smoothly with energy than the count-based maps.

Furthermore, the spectral inversion of Eq. (6.35) allows electron visibilities, and so electron flux maps, *to be determined at electron energies that are significantly greater than the maximum photon energy observed*. Physically, this is because photons at energy ϵ are produced by all electrons with energy $E > \epsilon$, so that the bremsstrahlung emission at a given energy contains significant information on electrons at all higher energies (Eq. (6.26)). The extent to which the electron visibility spectrum can be extended in this manner depends on both the quality of the count-based visibility data and the extent to which electrons of energy $E > \epsilon$ contribute to the bremsstrahlung emission at photon energy ϵ. This contribution decreases quite rapidly with E because of both the intrinsic steep decrease in the electron flux with energy ($F \sim E^{-\delta}$) and the decreasing value of the cross-section $Q(\epsilon, E)$. Simulations [e.g., 154] show that the electron visibilities can be reliably obtained up to energies $E \simeq 2\epsilon$. Figure 6.12 (after [191]) illustrates this smoothness with maps of the mean electron source spectrum for a flare on 2002 April 15, over a set of 2 keV energy bands spanning the range from 14–34 keV.

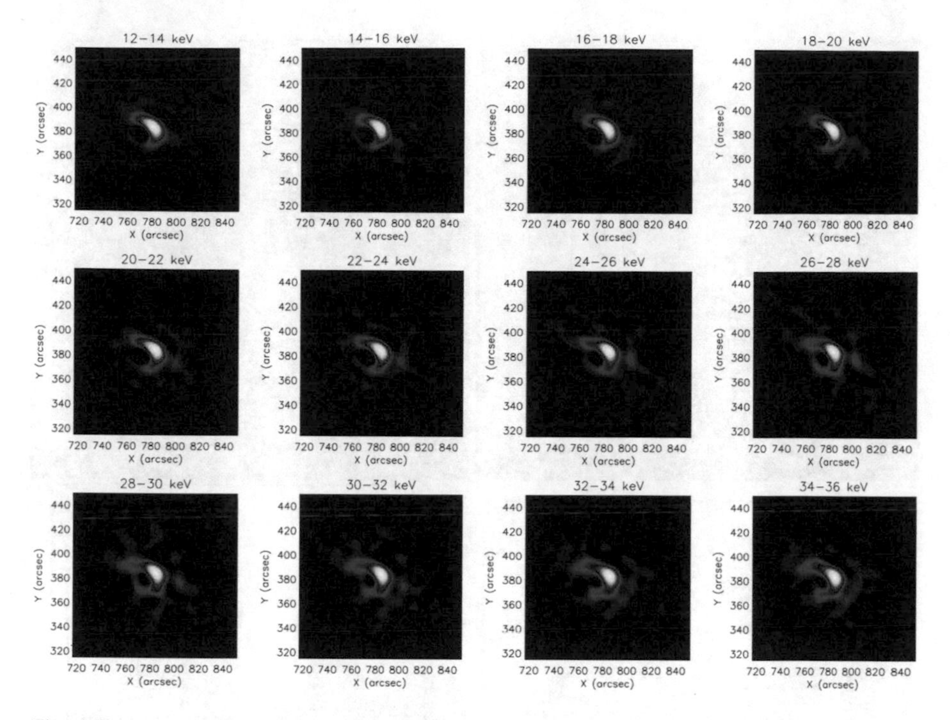

Fig. 6.12 Electron maps for the 2002 April 15 event. From left to right along each row, and from top to bottom, the images are for the energy bands 12–14 keV, 14–16 keV, $\cdots$, 34–36 keV, respectively. After [191],© AAS. Reproduced with permission.

The power of the electron visibility approach is illustrated [154] by its application to the C7.5 flare of 2002 February 20 (11:06:18 UT peak GOES flux). (This well-studied event has been discussed elsewhere in this book. In Sects. 5.3 and 5.4 the count-based forward-fit and Pixon algorithms, respectively, were used to model the two dominant footpoint sources on the left (East) side of the image; the overall morphology of the image, and in particular the likely number of predominant sources it contains at different count energies, is the subject of a Bayesian inference approach in Sect. 7.1.) Here, electron visibilities $W(u, v; E)$ were inferred from regularized inversion of the count-based visibility data using Eq. (6.32), and then the uv_smooth method (Sect. 6.6) was used to construct the corresponding electron flux images $F(x, y; E)$.

These electron flux maps are shown in Fig. 6.13. Four different spatial sub-regions in the source, labeled in that figure, were then selected [154] for analysis. Two of these regions correspond to the footpoint sources visible at higher energies (these sources are labeled "northern" and "southern" in Sect. 7.1) and the other two correspond to similarly-sized regions located approximately midway between the footpoints. The bottom left panel of Fig. 6.13 shows the electron flux spectra for each of these four sources, obtained by "stacking" the electron flux images at different energies. As predicted, because of the smoothing inherent in the spectral

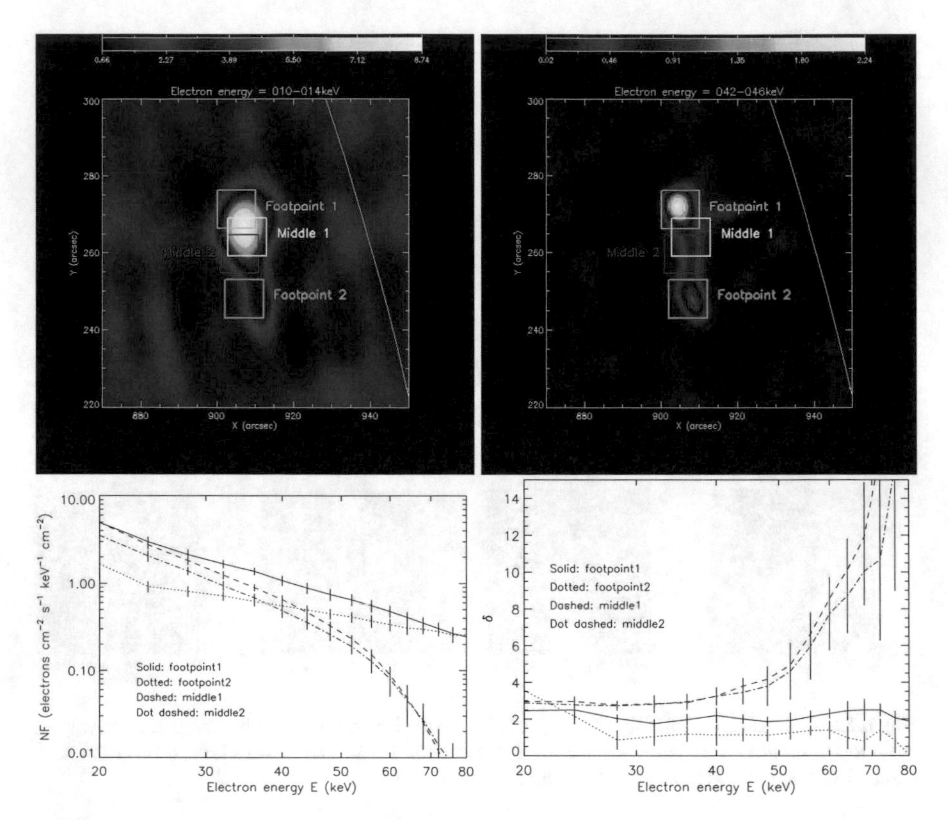

Fig. 6.13 *Top panels*: Electron images in the energy ranges 10–14 keV and 42–46 keV, respectively. Three sub-regions of interest are labeled on each image. Two of these correspond to bright footpoint-like sources and two to regions midway between the footpoints. *Bottom panel*: Electron spectra (left) and local spectral index (right) for each of the four sub-regions shown. After [154], © AAS. Reproduced with permission.

inversion (Eq. (6.32)) used to construct the electron visibilities, these electron flux spectra are much smoother than the photon spectra that could be recovered from photon maps.

The spectra corresponding to the two footpoint regions are clearly flatter (harder) than those corresponding to the regions between these footpoints. In order to quantify this spectral difference, the right panel of Fig. 6.13 shows the corresponding local spectral indices $\delta = -d \log F(E)/d \log E$ as a function of E, for each region. At electron energies $E \lesssim 60$ keV, the spectral index at the more southern footpoint (here labeled "Footpoint 2") is $\delta \simeq 2$, significantly smaller (i.e., the spectrum is flatter) than that at the more northern footpoint ("Footpoint 1")—see the left panel of Fig. 6.13. Above 60 keV the footpoints become equally bright (right panel of Fig. 6.13), with similar spectral indices. However, the spectrum of Footpoint 2 is very hard ($\delta \simeq 1$) and by $\sim$60 keV the electron fluxes at the two footpoints have become roughly equal, as is apparent from the spatial images. Above

$E \sim 40$ keV, the spectra corresponding to the two footpoint regions are significantly flatter (harder) than those corresponding to the regions between these footpoints.

Such a behavior is consistent with the physics of electron propagation in the ambient medium. Because the rate of energy loss due to Coulomb collisions is inversely proportional to energy (Eq. (1.17)), the lower-energy end of the electron spectrum depletes faster than the high-energy end, leading to a spectral flattening. At large distances from the acceleration region, the spectrum is significantly flatter (harder) than the spectrum that emerges from the acceleration region [24]. The average value of the difference $\Delta\delta$ in the range 40–70 keV in Fig. 6.13 is $\sim$2.0, consistent with a physical picture in which the electrons are accelerated in a source midway between the footpoints, and the subsequent propagation of these electrons to the footpoints flattens the spectrum. One can also account for the much harder spectrum of Footpoint 2 compared to Footpoint 1 (or, equivalently, the much weaker emission at low energies at Footpoint 1) if the column density distance between the acceleration region and Footpoint 2 is greater than that between the acceleration region and Footpoint 1 (see [62]). The higher flux at low energies in the "Middle 2" region suggests that, in fact, the acceleration region is closer to the "Middle 2" region and so to Footpoint 1, consistent with the above suggestion.

This interpretation of the observed spectral hardening requires a column density $N = \int n\, ds$ (cm^{-2}) $N < E^2/2K \simeq 2 \times 10^{17}$ [E(keV)]$^2 \simeq 3 \times 10^{20}$ cm^{-2} between the coronal acceleration region and the chromospheric footpoints. This in turn establishes an upper limit on the coronal density $n \sim N/d$, where d is the distance between the coronal source and the footpoint parallel to the guiding magnetic field. The plane-of-sky projected distance between the "middle" and "footpoint" sources in Fig. 6.13 is $\simeq 10'' \simeq 7 \times 10^8$ cm. Assuming a semicircular geometry for the loop connecting the footpoints with a length $d \simeq (\pi/2)$ times this projected distance, i.e., $\simeq 10^9$ cm, we infer that the coronal density $n \lesssim 3 \times 10^{11}$ cm^{-3}, an entirely reasonable value.

Chapter 7
Application to Solar Flares

Abstract In this chapter we apply the algorithms discussed in the previous chapters to selected solar flare events and discuss the physical implications of the results obtained. These investigations have a wide scope and include the following topics:

- The number and physical nature of hard X-ray sources in flares;
- Properties of the region in which the accelerated electrons are produced;
- Energy loss processes affecting accelerated electrons;
- Energetics of accelerated electrons and heated plasma and their contribution to the global energy budget of a solar eruptive event.

7.1 Number and Nature of Hard X-Ray Sources in the 2002 February 20 Event

The Bayesian optimization method (Sect. 6.4) has been applied [169] to a well-studied flare that occurred on 2002 February 20, only 2 weeks after *RHESSI*'s launch; this event has also been discussed in Sects. 5.3, 5.4, and 6.6 of this book. As discussed in these sections, and in the introduction to [169], the morphology of this flare had been a source of considerable debate, with a great deal of uncertainty regarding the identification of the "complicated pattern of low- and high-energy sources" [81] as either chromospheric footpoints or coronal sources.

Figure 7.1 shows the distribution of CLEAN point-source components (blue dots) and intensity contour levels obtained using the count-based CLEAN algorithm (Sect. 5.2) for the time interval from 11:06:10–11:06:24 UT. *RHESSI* detector #s 3 through 9 were used with the default CLEAN beam width and the three panels (left to right) correspond to count energy bands from 20–30 keV, 30–50 keV, and 50–70 keV, respectively. At the left of each image are two compact sources identified by Krucker and Lin [114] as loop footpoints extending over the full energy range. In Sect. 5.3, it was shown that the systematic shift of the footpoint locations with energy, determined by forward-fitting a double Gaussian source at each energy, is consistent with the greater penetration depth of higher-energy electrons.

M. Piana et al., *Hard X-Ray Imaging of Solar Flares*,
https://doi.org/10.1007/978-3-030-87277-9_7

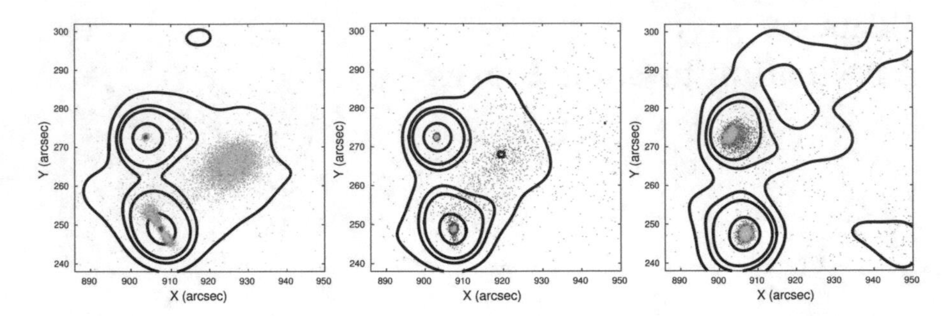

Fig. 7.1 Morphology of the 2002 February 20 flare at 11:06:10–11:06:24 UT in three count energy channels, 20–30, 30–50, and 50–70 keV, left to right. Each panel shows the distribution of the CLEAN point-source components in blue and the CLEAN contours in black at 8, 20, 30, and 70% of the peak intensity with the images made using *RHESSI* detectors 3 through 9 and the default CLEAN beam width. After [169], © AAS. Reproduced with permission.

There is also a third source to the right of the footpoint sources. It has lower intensity but nevertheless is clearly identified by the large number of CLEAN components in both the 20–30 and 30–50 keV images. This third source is less prominent in the higher energy channels than the footpoints, showing that its spectrum is significantly steeper. Such a spectrum is indicative [114] of a coronal "loop-top" source, which involves a combination of thermal [134] and non-thermal [116] emission in a relatively low-density environment.

Figure 7.2 shows the results of applying the Bayesian optimization algorithm to this event. The upper bar chart shows the posterior probabilities for the existence of $N = 1$ through 5 sources (colors) in the six count energy channels indicated. In the lowest (6–10 keV) energy channel, it is almost certain (83% probability) that there are two main sources present, the two "footpoint" sources on the left (Eastward) side of the CLEAN image (Fig. 7.1). At such low energies, the emission is most likely thermal in origin, and the predominance of the footpoint sources is presumably due to their higher density compared to the "loop-top" source in the corona.

In the next energy channel (10–14 keV), the source morphology with the highest likelihood now shifts toward three sources, consistent with the emergence of the source identified as "loop-top" at these energies. In the next energy channel (14–20 keV), the morphology with the highest probability (over 90%) is one with all three sources (two footpoints and a loop-top coronal source) present. The 20–30 keV channel is also characterized by those three predominant sources, although there is a not negligible probability that a fourth source existed. Identifying the location and size of such a source is still an open issue.

As we progress through the next highest energy channels (30–50 and 50–70 keV), the most likely source morphology returns to the two-component "double footpoint" structure, consistent with the CLEAN images of Fig. 7.1. The relative lack of brightness of the "loop-top" source at high energies is consistent both with the relative steepness of the exponential thermal (Maxwellian) spectra (Eq. (1.2))

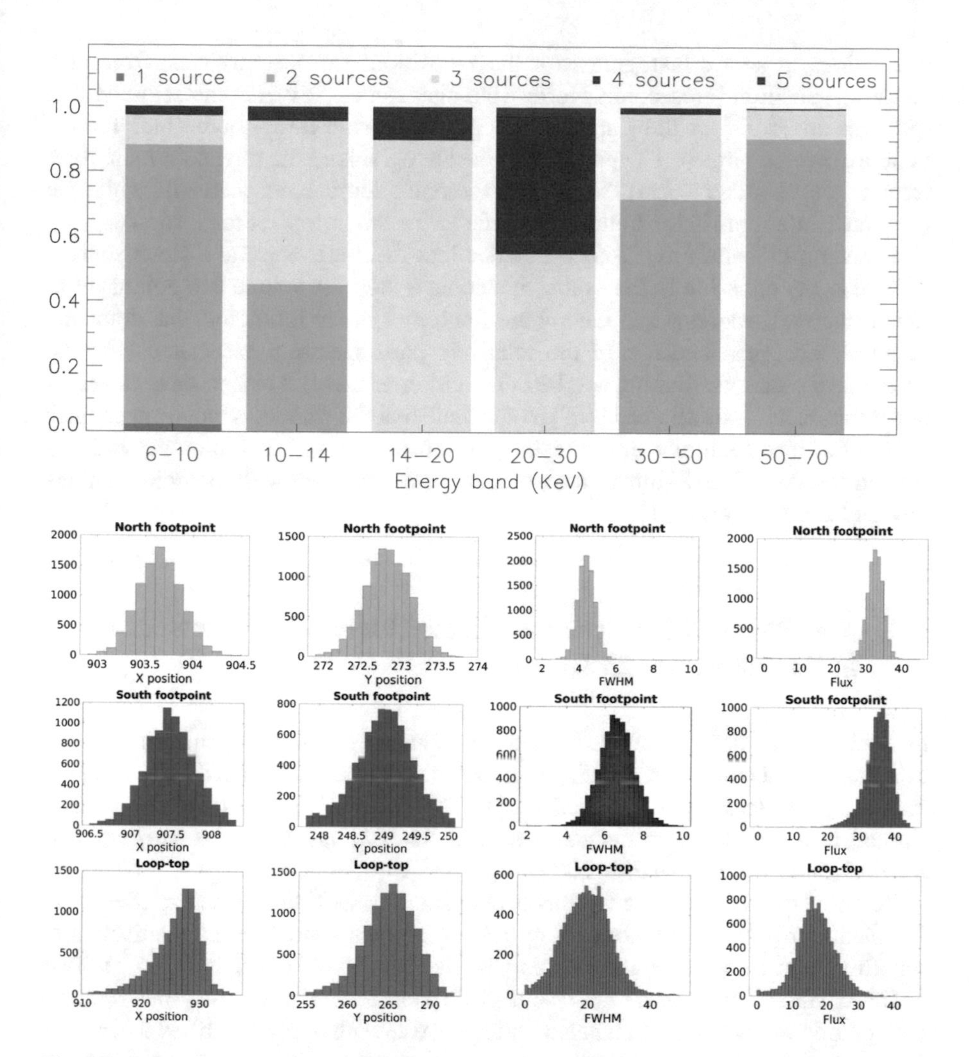

Fig. 7.2 *Top panel*: Bar chart showing the Bayesian conditional probabilities of N sources ($N = 1, 2, 3, 4, 5$; each assigned a different color as shown at the top of the Figure) for six different energy channels between 6 and 70 keV and the time interval 11:06:10–11:06:24 UT in the 2002 February 20 flare. *Bottom panel*: Posterior distributions for the X-position (arcseconds West of solar disk center), Y-position (arcseconds North of solar disk center), FWHM (arcseconds), and flux (photons cm^{-2} s^{-1} at the Earth) of the three dominant sources (two footpoints—"North" and "South"—and a "loop-top" coronal source) identified in the 20–30 keV image. After [169], © AAS. Reproduced with permission.

compared to power-law spectra, and with the relatively low column density of the "loop-top" source as a target for nonthermal emission from accelerated electrons.

The lower panel of Fig. 7.2 provides information on the characteristics of the emission in the mid-energy 20–30 keV channel. Based on 10,000 Monte Carlo realizations, it shows histograms for the x-position[1] (arcseconds West from disk center), y-position (arcseconds North from disk center), FWHM (arcsec) and flux (photons cm^{-2} s^{-1} at the Earth) of the most probable three-source morphology visible at such energies. The green and blue histograms in the top and middle rows refer to the "Northern" and "Southern" footpoint sources, respectively, while the magenta histograms in the bottom row refer to the "loop-top" coronal source.

Based on this analysis, [169] concluded that there are "several distinct sources of hard X-ray emission in the event," including a loop-top source that dominates at lower (thermal) energies and a number of chromospheric footpoints that dominate at higher energies (because of the relatively hard spectrum associated with the accelerated electrons that impact the dense chromosphere). Further, their Bayesian optimization analysis allowed [169] to conclude that the claimed sudden appearance [114] of a third footpoint in the vicinity of the "Southern" footpoint was rather a shift in position of the Southern footpoint at a rate consistent with footpoint motions reported in other events [115].

7.2 The Physical Nature of Multiple Hard X-Ray Sources in the 2002 July 23 Event

In Sect. 6.6, we demonstrated the general robustness of the uv_smooth method, using simulated data corresponding to a two-dimensional Gaussian source. In [133] the uv_smooth technique was applied to the 2002 July 23 flare, an event previously studied in a number of papers in a special *RHESSI* issue of *The Astrophysical Journal (Letters)* (volume **595**, 2003).

Analysis of *RHESSI* data for this event using the CLEAN algorithm (Sect. 5.2) revealed four prominent features (Fig. 7.3)—a coronal source with a relatively soft spectrum, and three compact sources (see the 36–41 keV panel in Fig. 7.3). Two of these, labeled "northern" and "southern," were identified as footpoint sources, and a third compact source, labeled "middle," was tentatively identified as another footpoint on the basis of context information at other wavelengths, such as Hα [62].

Figure 7.4 shows a comparison of the uv_smooth (Sect. 6.6) results (bottom panel in figure) with results obtained with the natural-weighting CLEAN algorithm (top panel). While the overall morphology of the flare is confirmed in all sets of images, there are some significant differences in the uv_smooth images:

1. Overall, the uv_smooth method generates images with very similar morphology to those produced by the natural-weighting CLEAN algorithm. However,

[1] Recall that 1 arcsecond ($1''$) corresponds to $\sim$725 km on the Sun, when viewed from the Earth at a normal (i.e., 90°) angle to the solar surface; a $\sec\theta$ de-projection factor must be applied in calculating distances on the solar surface for sources at a heliocentric angle θ from disk center. The heliocentric angle for a source with coordinates (x, y) is given by $\sin\theta = \sqrt{x^2 + y^2}/R$, where $R \simeq$960 arcseconds is the solar radius as viewed from the Earth.

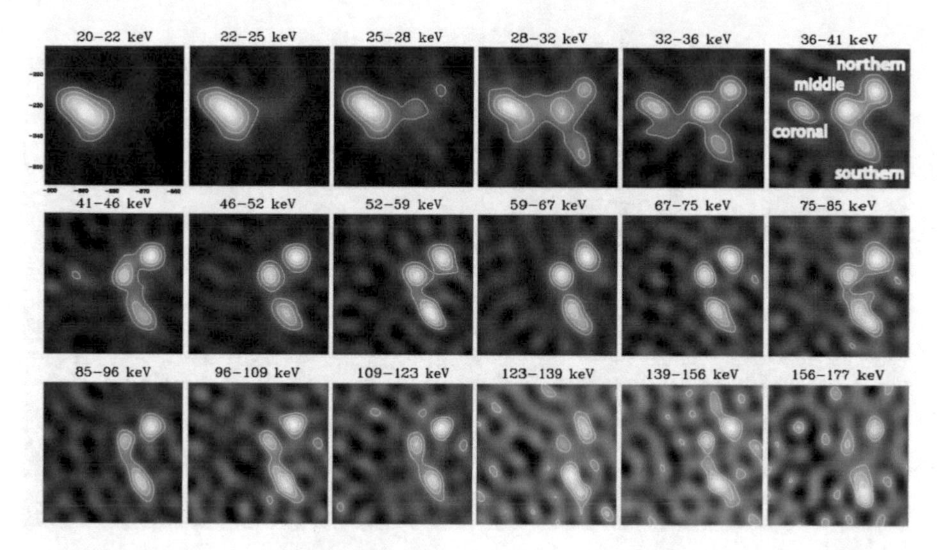

Fig. 7.3 Images obtained for the 2002 July 23 event, using the CLEAN algorithm with *uniform* detector weighting and the default clean beam width. The contour levels are set at 30, 50, 70, and 90% of the peak intensity. The flare was located close to the East limb, so that distances to the left represent mostly vertical height. Four principal sources, labeled "northern," "southern," "middle," and "coronal" are identified. After [62], © AAS. Reproduced with permission.

because uv_smooth uses information provided by Fourier components interpolated between observed (u, v) values, features produced are generally sharper than those produced by CLEAN (Fig. 7.4). However, this difference is reduced if a clean beam width of about half the default value is used.

2. Both the CLEAN and uv_smooth images (Fig. 7.4) show emission extending along a curved locus joining the "northern" footpoint with the "southern" one. This suggests that the "middle" source is not a footpoint, but rather a coronal "loop-top" source containing material that has evaporated up along a loop in response to flare heating at the footpoints (see, e.g., [128]).
3. The "coronal" source, visible only at energies up to and including the 36–41 keV channel in the uniform-weight CLEAN images (Fig. 7.3), is still visible all the way up to 85 keV in both the natural-weight CLEAN and uv_smooth images (Fig. 7.4). Its absence in the uniform-weight CLEAN images at energies above 41 keV is a consequence of both sampling and dynamic range issues—at higher count energies, the footpoints have a higher intensity per pixel and the uniform-weight CLEAN algorithm, with its reduced weighting of (u, v) points corresponding to smaller spatial scales, simply ignores the relatively weak coronal source in its synthesis of the image through systematic assembly of the brightest pixels in the field of view. By contrast, using natural weights in the CLEAN algorithm, or using uv_smooth, equal weight is placed on the information from all sampled visibilities, including those at large radii in the (u, v) plane (Fig. 6.1). Such a data sampling preserves information on *all* compact

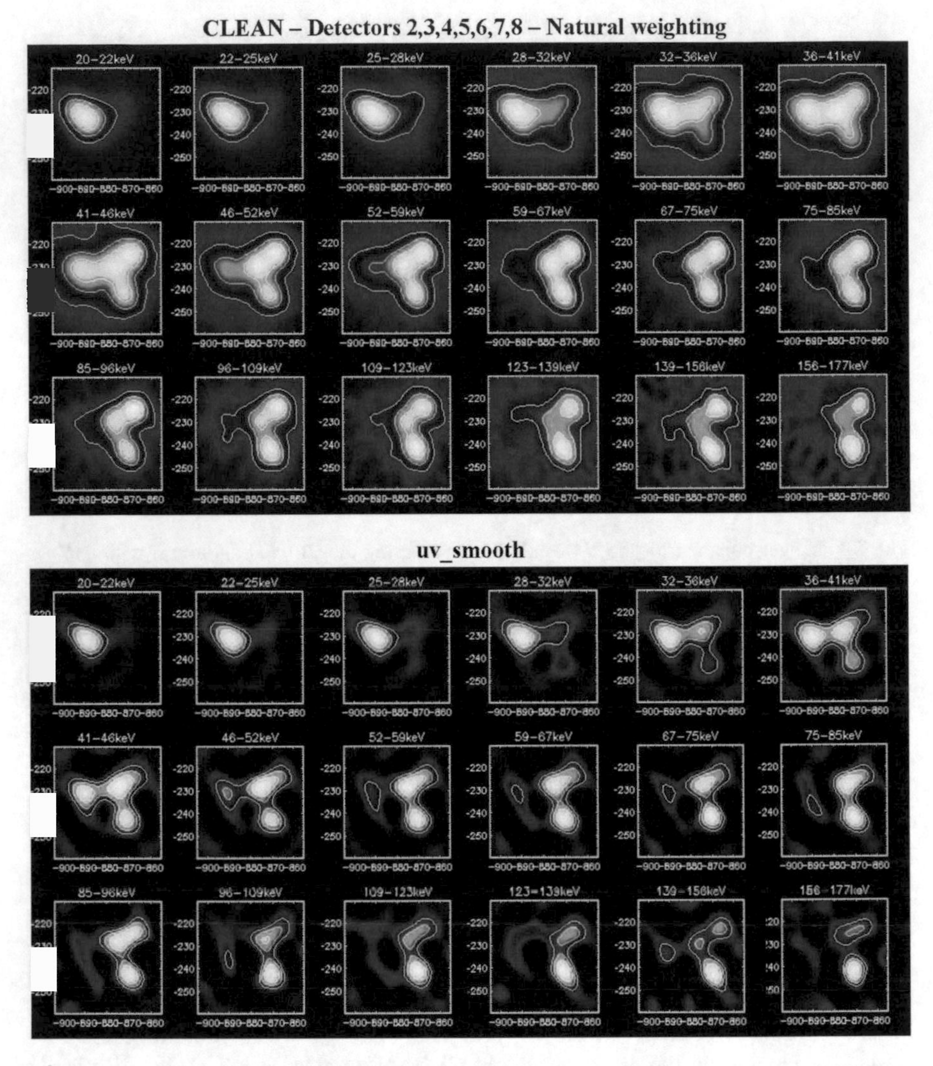

Fig. 7.4 Comparison of the images obtained for the 2002 July 23 event, using both the CLEAN algorithm with *natural* detector weighting and the default clean beam width (top panels), and uv_smooth (bottom panels). After [133], © AAS. Reproduced with permission.

sources in the field of view, so that using natural weights in CLEAN, or using uv_smooth, continues to map the coronal source even when its brightness per unit area is much smaller than that of the compact footpoints.

The appearance of the curved locus joining the "northern" and "southern" features suggests that they are both elements of a single extended source. In [62], it was shown that there is a small, but significant, difference between the hard X-ray spectra of the "northern" and "southern" sources, with the spectrum of the "southern" source being somewhat flatter (i.e., harder) than that of the "northern" source. Given that the electron energy loss rate due to Coulomb collisions is inversely proportional to energy (Eq. (1.17)), collisions act to preferentially reduce the number of low-energy electrons and so cause a hardening of the spectrum of the injected electrons. As pointed out in Appendix A of [162], the intervening column density necessary to accomplish the measured degree of spectral hardening requires that substantial hard X-ray emission be produced in the intervening material, consistent with the appearance of the "bridge" of emission connecting the "northern" and "southern" sources.

7.3 Properties of the Electron Acceleration Region

7.3.1 Using the VIS_FWDFIT Method to Estimate the Acceleration Region Length and Density

In this section, we illustrate the use of the VIS_FWDFIT method (Sect. 6.3) to determine general properties of the X-ray sources. One of the most useful properties is the variation of source size with photon energy since this can help to constrain scientifically important physical parameters, such as the density in the hard X-ray target and the extent of the electron acceleration region. However, we then present a cautionary note that by its very nature, results using VIS_FWDFIT are only as reliable as the assumptions that are used in making such a fit. In particular, parametric fitting of an image using an overly simplistic source geometry can result in apparent trends in the source parameters that may in fact result from a source with a geometry that is not consistent with the simple form assumed.

The visibility-forward-fit method was first applied [206] to a set of nine solar flare events, each of which appeared to have a relative simple extended-coronal-loop geometry in the energy range of interest. Figure 7.5 compares the CLEAN and VIS_FWDFIT images, showing reasonable agreement with such an assumed geometry. (However, it should be noted that the CLEAN images in higher energy ranges do suggest significant departures from the assumed geometry for some of the events.) The second and fourth rows of Fig. 7.5 show the visibility-forward-fit images in the representative energy ranges 10–15 keV, and 15–30 keV, respectively. Careful examination of these images shows that the size of the source, particularly along the longest dimension, increases with increasing energy.

Based on their appearance in Fig. 7.5, these sources were assumed to have the curved elliptical Gaussian form shown in Fig. 6.3 and represented by the following equation (identical to Eq. 6.1):

$$I(x, y; \epsilon) = I_o(\epsilon)\, e^{-s^2/2\sigma^2}\, e^{-t^2/2\tau^2} , \tag{7.1}$$

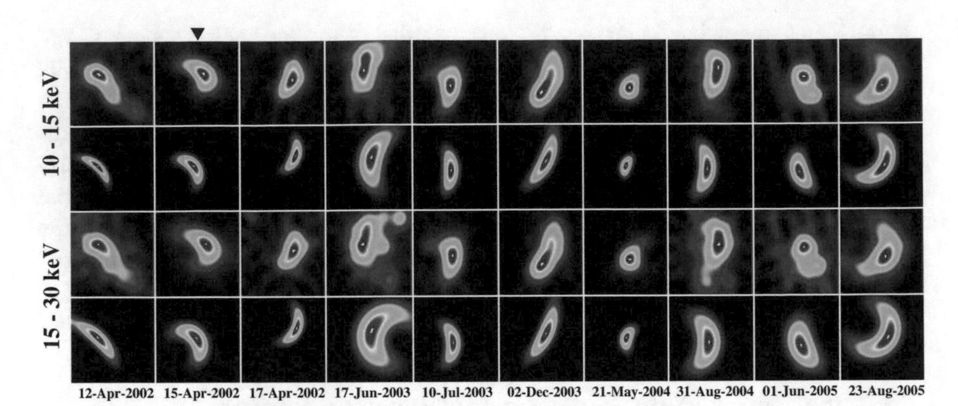

Fig. 7.5 CLEAN and VIS_FWDFIT images for the events in the visibility-forward-fit study of [206]. First and third rows: CLEAN images (made with the default clean beam width) in the energy ranges 10–15 and 15–30 keV, respectively. Second and fourth rows: VIS_FWDFIT images for the same energy ranges made assuming a single curved elliptical Gaussian in each case. After [206], © AAS. Reproduced with permission.

where $s(x, y)$ is a coordinate along a circular arc of radius R centered on a point (x_o, y_o) on the plane of the sky, with the line from (x_o, y_o) to the center of the arc (with intensity I_o) making an angle θ_o relative to a reference direction (e.g., North), $t(x, y)$ is the coordinate locally perpendicular to this arc, and $\sigma(\epsilon)$ and $\tau(\epsilon)$ are the (energy-dependent) standard deviations of the source extent in the parallel and perpendicular directions, respectively. This spatial form, and hence its (energy-dependent) Fourier transform

$$V(u, v; \epsilon) = \int\int I(x, y; \epsilon) \exp\left(2\pi i[u(x - x_o) + v(y - y_o)]\right) , \tag{7.2}$$

are both functions of the seven parameters $(I_o, x_o, y_o, \theta_o, R, \sigma, \tau)$. As discussed in Sect. 6.3, the VIS_FWDFIT routine finds the values of these seven parameters that provide the best χ^2 fit to the observed visibilities $V(u, v)$ at the (u, v) points sampled by the instrument.

For analytic convenience (see discussion below), the (energy-dependent) extent of the source along its guiding arc was modeled in [206] not in terms of the standard deviation $\sigma(\epsilon)$ but rather in terms of the "one-sided centroid" quantity s_c, corresponding to the centroid position along the s-direction for one half of the (symmetrical) structure:

$$s_c(\epsilon) = \frac{\int_0^\infty s\, I(\epsilon, s)\, ds}{\int_0^\infty I(\epsilon, s)\, ds} . \tag{7.3}$$

Using the Gaussian form (7.1) in Eq. (7.3) gives

$$s_c = \frac{\int_0^\infty s\, e^{-s^2/2\sigma^2}\, ds}{\int_0^\infty e^{-s^2/2\sigma^2}\, ds} = \sqrt{\frac{2}{\pi}}\, \sigma \,, \tag{7.4}$$

so that s_c is proportional to the fit parameter σ.

The reason for the choice (7.3) of s_c to measure the longitudinal extent of the source is for analytical convenience, as we now show. In a collisional medium, the electron energy varies with distance according to Eq. (1.17):

$$\frac{dE}{ds} = -\frac{Kn}{E} \,, \tag{7.5}$$

where E (keV) is the electron energy, s (cm) is position along the electron trajectory, n (cm^{-3}) is the ambient particle density and $K = 2.8 \times 10^{-18}\,\mathrm{cm^2\,keV^2}$ is a constant defining the strength of the Coulomb interaction process [58]. The solution of Eq. (7.5) is

$$E^2 = E_o^2 - 2Kns \,, \tag{7.6}$$

where E_o is the energy of the electron when it leaves the acceleration region. The electron flux continuity equation

$$F(E, s)\, dE = F_o(E_o)\, dE_o \,, \tag{7.7}$$

combined with the result $E\, dE = E_o\, dE_o$ derived from (7.6), gives the electron flux (cm^{-2} s^{-1} keV^{-1}; differential in energy E) as a function of E and s:

$$F(E, s) = \frac{E}{\sqrt{E^2 + 2Kns}}\, F_o(\sqrt{E^2 + 2Kns}) = A\, E\, (E^2 + 2Kns)^{-(\delta+1)/2} \,, \tag{7.8}$$

where we have assumed a power-law form $F_o(E_o) = AE_o^{-\delta}$. We can now use Eq. (1.8) relating the bremsstrahlung X-ray spectrum at Earth to the spectrum of the emitting electrons at the Sun to determine the hard X-ray yield as a function of photon energy and position along the loop:

$$I(\epsilon, s) = n \int_\epsilon^\infty F(E, s)\, Q(\epsilon, E)\, dE = \frac{n\, Q_o\, A}{\epsilon} \int_\epsilon^\infty (E^2 + 2Kns)^{-(\delta+1)/2}\, dE \,, \tag{7.9}$$

where we have for simplicity used the Kramers cross-section $Q(\epsilon, E) = Q_o/\epsilon E$ (Eq. (1.9)). While the standard deviation $\sigma(\epsilon)$ of such an (I, ϵ, s) form is rather difficult to calculate, evaluation of the one-sided centroid s_c, Eq. (7.3), is relatively straightforward, thus explaining its use. Using Eq. (7.9) in the defining Eq. (7.3) gives

$$s_c(\epsilon) = L + \frac{\int_0^\infty s \int_\epsilon^\infty (E^2 + 2Kns)^{-(\delta+1)/2}\, dE\, ds}{\int_0^\infty \int_\epsilon^\infty (E^2 + 2Kns)^{-(\delta+1)/2}\, dE\, ds} , \tag{7.10}$$

where we have added a distance L, corresponding to the half-width of the acceleration region at the top of the loop from which the electrons emerge (so that $s = 0$ corresponds to a distance L from the apex of the loop). Reversing the order of integration and changing the variable to $x = E^2 + 2Kns$,

$$\begin{aligned} s_c(\epsilon) &= L + \frac{1}{2Kn} \frac{\int_\epsilon^\infty dE \int_{E^2}^\infty (x - E^2)\, x^{-(\delta+1)/2}\, dx}{\int_\epsilon^\infty dE \int_{E^2}^\infty x^{-(\delta+1)/2}\, dx} \\ &= L + \frac{1}{2Kn} \frac{\frac{2}{(\delta-3)} \int_\epsilon^\infty E^{3-\delta}\, dE \; - \; \frac{2}{(\delta-1)} \int_\epsilon^\infty E^2\, E^{1-\delta}\, dE}{\frac{2}{(\delta-1)} \int_\epsilon^\infty E^{1-\delta}\, dE} \\ &= L + \frac{(\delta - 2)}{(\delta - 3)(\delta - 4)} \times \frac{\epsilon^2}{Kn} . \end{aligned} \tag{7.11}$$

Equation (7.11) shows that images at larger photon energies should have one-sided centroids that are farther away from the center of the source, with the variation being quadratic in photon energy ϵ. Physically, this is because higher-energy electrons have a penetration column depth that scales as the square of the electron energy E.

The top two rows of Fig. 7.6 shows the variation of s_c with photon energy ϵ, fit to the quadratic relation (7.11) in order to determine the density n and half-length L of the acceleration regions. These two quantities vary from event to event, over the ranges $\sim 2 \times 10^{11}$ cm^{-3} to $\sim 4 \times 10^{11}$ cm^{-3}, and $\sim 2''$ to $\sim 6''$, respectively.

7.3.2 Using the MEM_NJIT Method to Revisit Earlier Results

The above conclusions depend crucially on the finding that the length of the coronal hard X-ray source increased with energy in the 20–30 keV range. However, as pointed out above, the CLEAN images in higher energy ranges of Fig. 7.5 indicate a departure from the assumed simple geometry required for the VIS_FWDFIT method. Consequently, this conclusion was revisited in [52], where a more sophisticated image reconstruction technique was used, based on the MEM_NJIT algorithm (Sect. 6.5). There it was shown that footpoint emission at higher energies affects the inferred length of the coronal hard X-ray source when just a single curved Gaussian source is assumed.

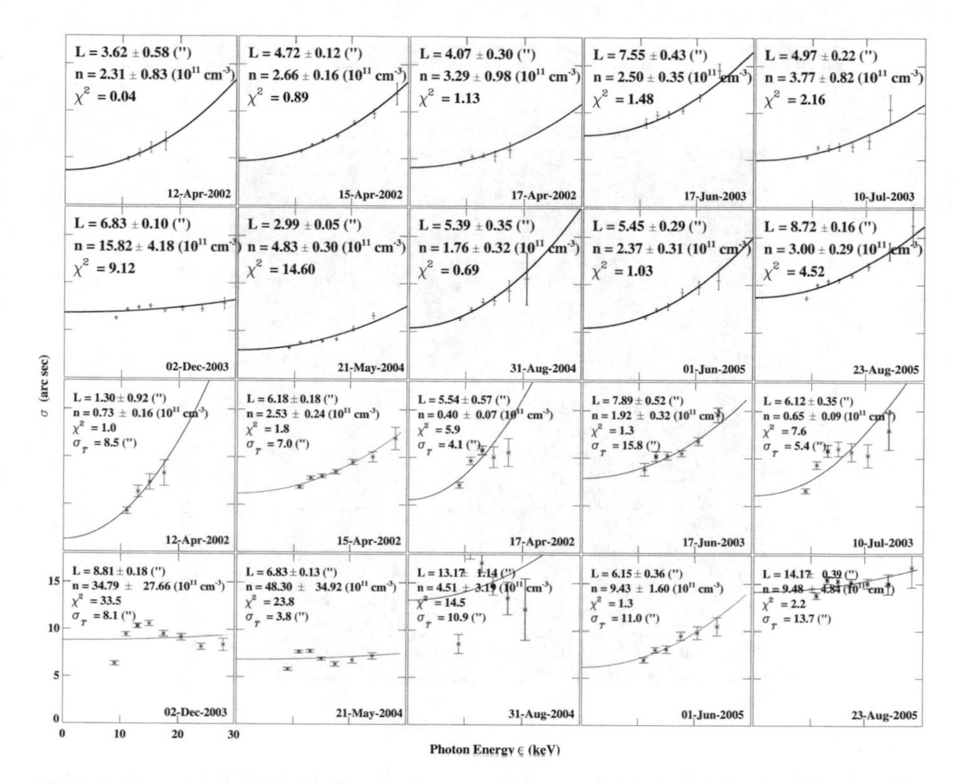

Fig. 7.6 The top two rows show plots of the one-sided centroid $s_c(\epsilon)$ (Eq. (7.3)) for each of the ten events, together with the best fit lines of the form $L + (b/n)\,\epsilon^2$ (cf. Eq. (7.11)). The best-fit values of L (arc seconds) and n (cm^{-3}) are given. The bottom two rows show fits to a different model, in which energy losses in the acceleration region itself are taken into account. After [206], © AAS. Reproduced with permission.

Instead of assuming that the source was a single elliptical Gaussian, [52] allowed the image to contain three separate sources—an extended coronal source and two compact footpoints. This morphology is illustrated in Fig. 7.7, where the 12–25 keV image made with MEM_NJIT shows a single extended source but the 25–50 keV overlaid contours reveal two compact sources at each end of the extended source. Since the spectrum of the footpoint sources is harder than that of the coronal source, the footpoint sources make an increasing contribution to the total emission with increasing energy. Hence, if a single elliptical Gaussian were used to fit the data, as was done in [206], then one would incorrectly conclude that the source increased in length with increasing energy.

Based on this new analysis, it was concluded [52] that the lengths of the hard X-ray coronal sources do *not* systematically increase with energy in any of the events analyzed. In fact, for the six flares and 12 time intervals that satisfied the selection criteria, the loop lengths *decreased* on average by $(1.0 \pm 0.2)''$ between 20 and 30 keV. They did, however, find strong evidence that the *altitude* of the

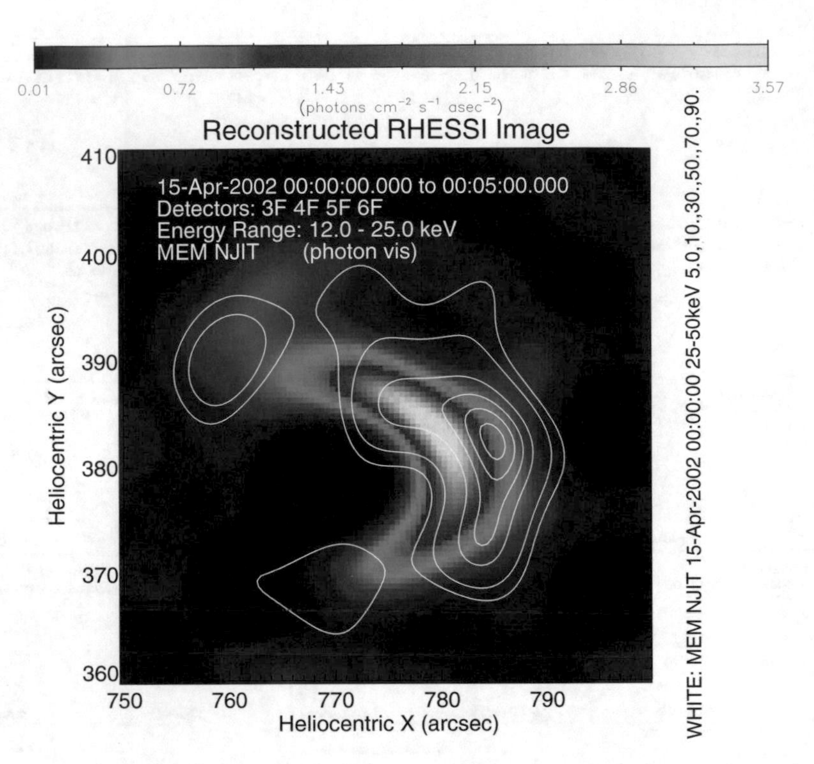

Fig. 7.7 *RHESSI* 12–25 keV image (color-coded according to the color bar, in units of photons cm^{-2} s^{-1} arcsec^{-2}, overlaid with 25–50 keV contours (at 5, 10, 30, 50, 70, 90% of the peak value) for the 5-min interval from 00:00 to 00:05 UT. The images were made using the MEM_NJIT reconstruction technique with data from the front segments of detectors 3, 4, 5, and 6. Two compact footpoints are evident in the 25–50 keV image contours. The extended coronal source is also present at both energies but is located further to the right in the higher energy image, and hence, most probably at a higher altitude. After [52], © AAS. Reproduced with permission.

intensity peak in the coronal hard X-ray source increases with increasing energy, as shown in Fig. 7.7. For the thermal component of the emission, this is consistent with the standard Carmichael-Hirayama-Sturrock-Kopp-Pneuman [CHSKP; 32, 88, 112, 182] flare model illustrated in Fig. 1.1 in which magnetic reconnection in a coronal current sheet results in new hot loops being formed at progressively higher altitudes.

This revision of an earlier result highlights the issues involved in using simple assumptions about the source structure. While VIS_FWDFIT can be a powerful tool for deriving source dimensions, spectra, and time variability, like all methods it must be used with appropriate caution.

7.4 Empirical Determination of the Electron Energy Loss Rate

In Sect. 1.2.3, we illustrated how to obtain the electron injection rate $\dot{N}_o(E_o)$ ($\mathrm{s}^{-1}\,\mathrm{keV}^{-1}$) from the mean source electron spectrum $\langle nVF(E)\rangle$, a quantity that is determined from observations of the hard X-ray intensity $I(\epsilon)$. There, specifically in Eq. (1.17), we assumed that the electron energy loss rate was given by that appropriate to electron-electron Coulomb collisions. However, other forms of energy loss may be occurring, such as wave-particle interactions and return-current effects, must be considered. Thus, it would be preferable to not *assume* the form of the energy loss rate, but rather use observations of the variation of the mean source electron spectrum $\langle nVF(E)\rangle$ at different locations in order to *determine* both the injected electron spectrum and the form of the energy loss rate that controls the spatial variation of $\langle nVF(E)\rangle$ throughout the source.

Determination of the variation of the mean source electron spectrum $\langle nVF(E)\rangle$ in separate regions of the flare volume is best done using the "electron visibility" method described in Sect. 6.9. To briefly recap, the essence of this method is to perform a *spectral* inversion of the count-based visibilities i.e., to solve the integral equation

$$V(u, v; q) = \frac{1}{4\pi R^2} \int_q^\infty W(u, v; E)\, K(q, E)\, dE \tag{7.12}$$

to obtain the electron visibilities $W(u, v; E)$ from the count-based visibilities $V(u, v; q)$. Once the electron visibilities have been obtained in this manner, the various visibility-based image reconstruction methods described in Chap. 6 may be used to convert them to a spatial map of the mean source electron flux spectrum $\langle nVF(E)\rangle(x, y)$. We can then appeal to the equation of continuity for electron flux [61] to deduce the form of the energy loss rate that is consistent with the variation of the mean source electron spectra from point to point.

We follow the analysis in [191], where s denote a spatial coordinate and $F(E, s)$ denotes the local electron flux spectrum (electrons $\mathrm{cm}^{-2}\,\mathrm{s}^{-1}\,\mathrm{keV}^{-1}$). The continuity of electron flux requires that

$$\frac{\partial F(E, s)}{\partial s} + \frac{\partial}{\partial E}\left(F(E, s)\,\frac{dE}{ds}\right) = S(E, s)\,, \tag{7.13}$$

where dE/ds is the electron energy loss per unit distance and $S(E, s)$ is a source term. Note that dE/ds is, in general, a function of both the electron energy E and the ambient conditions (e.g., density).

It must be noted that the observed X-ray flux is proportional to the mean source electron flux spectrum $\langle nVF(E)\rangle$, which itself is proportional to both the electron flux $F(E, s)$ and the number of target particles $n(s)V(s)$ in the observed sub-region. Therefor, the electron flux $F(E, s)$ cannot be directly inferred from the

X-ray observations. To circumvent this obstacle, it was assumed in [191] that $F(x, y, z; E) = \overline{F}(x, y; E)$ (i.e., that the electron flux spectrum is independent of position along the line-of-sight direction z), and the additional condition

$$\left| \frac{dN(s)}{ds} F(E, s) \right| \ll \left| N(s) \frac{\partial F(E, s)}{\partial s} \right| . \tag{7.14}$$

was also imposed. This condition is equivalent to

$$\left| \frac{d \ln N(s)}{ds} \right| \ll \left| \frac{\partial \ln F(E, s)}{\partial s} \right| ; \tag{7.15}$$

it essentially requires that the spatial variation of the electron spectrum $F(E, s)$ is more significant than the spatial variation of the line-of-sight column density $N(s)$ in determining the variation of the product $N(s) F(E, s)$. As discussed in [191], this assumption is valid for the electron energies and solar atmospheric conditions applicable.

The inequality (7.14) means that

$$\frac{\partial [N(s) F(E, s)]}{\partial s} \simeq N(s) \frac{\partial F(E, s)}{\partial s} , \tag{7.16}$$

and, with these assumptions, Eq. (7.13) can be transformed into

$$\frac{\partial g(E, s)}{\partial s} + \frac{\partial}{\partial E} \left(g(E, s) \frac{dE}{ds} \right) = h(E, s) , \tag{7.17}$$

where $g(E, s) = N(s) \overline{F}(E, s)$ (electrons cm^{-4} s^{-1} keV^{-1}) is the quantity represented in observationally-deduced electron maps and $h(E, s) = N(s) S(E, s)$ (electrons cm^{-5} s^{-1} keV^{-1}) is the line-of-sight-column-density-weighted source term.

The formal solution of Eq. (7.17) is

$$\frac{dE}{ds}(E, s) = \frac{1}{g(E, s)} \int_E^{\infty} \left[\frac{\partial g(E, s)}{\partial s} - h(E, s) \right] dE , \tag{7.18}$$

which is an empirical formula for the energy loss rate (a function of both electron energy E and position s), deduced from the observationally-inferred quantity $g(E, s)$, given the source term $h(E, s)$. (In deriving this equation, [191] assumed that the energy loss process is such that energy loss is negligible at high energies: $dE/ds \to 0$ as $E \to \infty$.)

Equation (7.18) may be rearranged as

$$\frac{dE}{ds}(E,s) + \frac{1}{g(E,s)} \int_E^\infty h(E,s)\,dE = \frac{1}{g(E,s)} \int_E^\infty \frac{\partial g(E,s)}{\partial s}\,dE \equiv -R(E,s)\,. \tag{7.19}$$

The quantity $R(E,s) = -(1/g(E,s)) \int_E^\infty (\partial g(E,s)/\partial s)\,dE$ can be directly inferred from observations with no physical assumptions required. In the absence of the injection of fresh electrons along the source ($h(E,s) = 0$), it represents the absolute value of the energy loss rate dE/ds. However, the injection of fresh electrons (given by $h(E,s)$) changes the electron energy loss rate that is required for consistency with the observationally-inferred variation of $g(E,s)$.

To evaluate $R(E,s)$, we first need to identify the meaning of the spatial coordinate s, the direction along the magnetic field lines that guide the accelerated electrons. Figure 7.8 shows the locus of points that define the main (toroidal) axis of

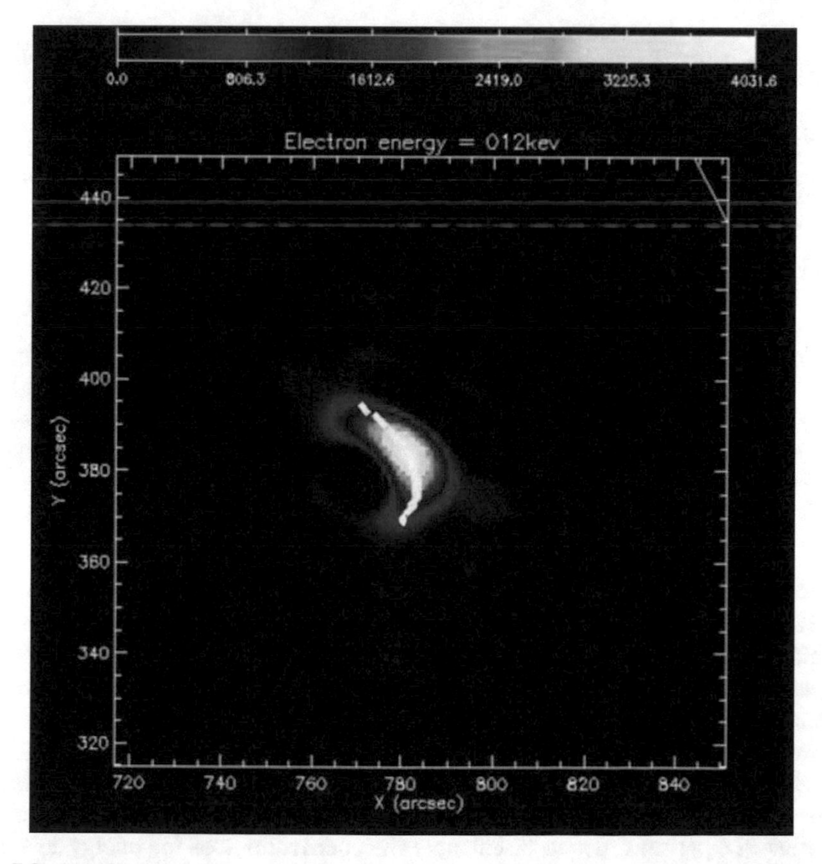

Fig. 7.8 Output of the image processing procedure that provides points that specify the direction of the guiding magnetic field axis of the loop. Here we have superimposed such points as a white line on the 12–14 keV image. After [191], © AAS. Reproduced with permission.

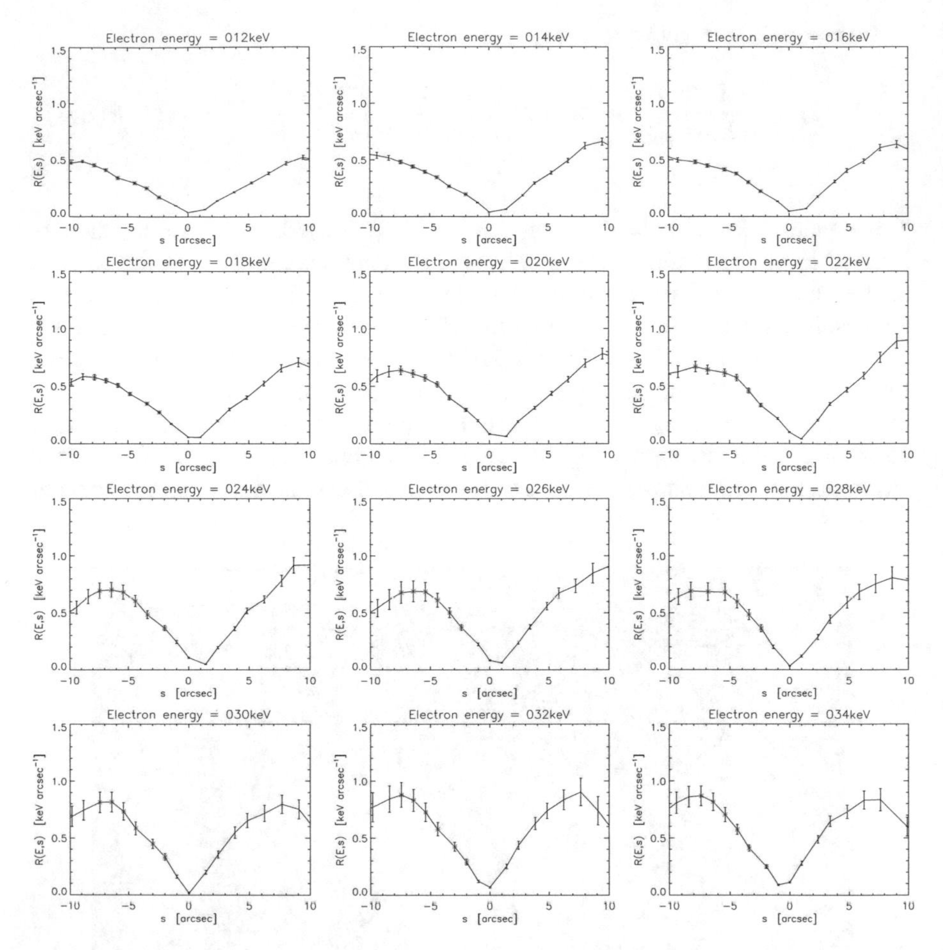

Fig. 7.9 $R(E, s)$ (see Eq. (7.19)) as a function of position s in the magnetic guiding line for the same energy channels used to construct the electron maps shown in Fig. 6.12. The six points used to compute the values of $R(E)$ plotted in Fig. 7.10 are indicated with asterisks in each frame. After [191], © AAS. Reproduced with permission.

the loop, i.e., the direction of coordinate s, in a 12–14 keV image of a flare observed on 2002 April 15 (00:05:00 UT). Using electron flux images at different energies, obtained using the electron visibilities deduced from inversion of Eq. (7.12) and the uv_smooth method, Fig. 7.9 shows the behavior of the quantity $R(E, s)$, plotted as a function of position s for different energies E (separate panels). Finally, Fig. 7.10 shows the average value of $R(E, s)$ for a set of points highlighted in Fig. 7.9.

As discussed in [191], the form of $R(E, s)$ obtained from this analysis is not consistent with plausible energy loss forms unless there is a non-zero source function $h(E, s)$. In order to disentangle the contributions of the energy loss rate

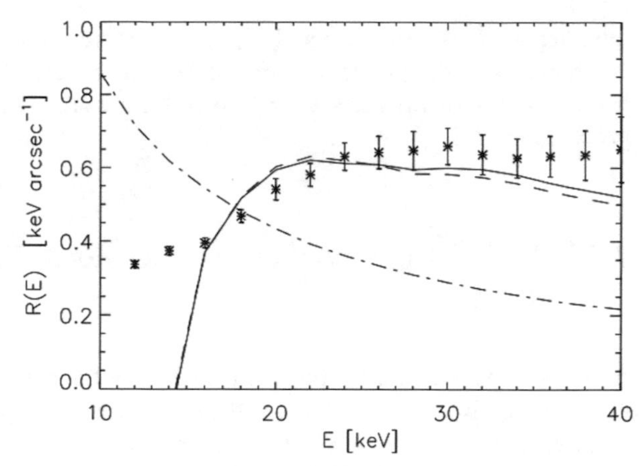

Fig. 7.10 $R(E)$, averaged over the six positions highlighted in Fig. 7.9, as a function of electron energy E. The dot-dashed, dashed, and solid lines correspond to fits of the theoretical model of Eq. (7.19) to the data, with temperature $T = 1.61$ keV and acceleration region length $L_o = 0, 10''$, and $16''$, respectively. After [191], © AAS. Reproduced with permission.

dE/ds and this source term $h(E, s)$, [191] adopted a parametric form for $h(E, s)$, namely

$$h(E, s) = \begin{cases} h_o \left(\frac{E}{E_o} \right)^{-\delta} & ; \ |s| \le \frac{L_o}{2} \\ 0 & ; \ |s| > \frac{L_o}{2} \end{cases} , \tag{7.20}$$

where δ is the spectral index of the injected electron spectrum and L_o is the half-length of the acceleration region. Fitting of the empirical form of the spatially averaged quantity $R(E)$ thus yields values of the acceleration region length L_o, the target density n, and the electron injection rate per unit volume.

The best-fit values obtained [191] were $L_o \simeq 1.1 \times 10^9$ cm, comparable to the value obtained [82] in an independent analysis of the same event, and $n \lesssim 2 \times 10^{11}$ cm^{-3}, consistent with estimates of the density obtained from considerations of the source emission measure: $n = \sqrt{EM/V} \simeq \sqrt{EM/A^{3/2}}$. The best-fit magnitude of the source term h_o (using a reference energy $E_o = 20$ keV in Eq. (7.20)) was $h_o = (1.4 \pm 0.1) \times 10^{28}$ cm^{-5} keV^{-1} s^{-1}. Integrating equation (7.20) from E_o upwards (using $\delta \simeq 10$; [191]) gives $h_o E_o/(\delta - 1) \simeq 3 \times 10^{28}$ cm^{-5} s^{-1}; this is the rate of acceleration of electrons per unit volume (cm^{-3} s^{-1}) to energies greater than 20 keV, weighted by the line-of-sight column density N (cm^{-2}).

The line-of-sight column density N can be estimated from EM $= n^2 V = n^2 A L_o = n N A$, where EM $\simeq 8 \times 10^{48}$ cm^{-3} is the emission measure of the soft X-ray source and A is the cross-sectional area of the loop, estimated from the images to be $\simeq 2.5 \times 10^{18}$ cm^2. With $n \simeq 2 \times 10^{11}$ cm^{-3}, this gives $N \simeq 1.5 \times 10^{19}$ cm^{-2} and so a volumetric source term equal to 3×10^{28} cm^{-5} s^{-1}/1.5×10^{19} cm^{-2} $\simeq$ 2×10^9 cm^{-3} s^{-1} above 20 keV. Dividing this by the ambient number density n (cm^{-3}) gives [191] a measure of the *specific acceleration rate* $\eta(>20\,\text{keV}) \simeq 10^{-2}$ electron s^{-1} per ambient electron. This important quantity represents the fraction of available electrons that are accelerated per unit time to energies greater than the 20 keV reference energy, and is well within the range of specific acceleration rates reported for other events [65]. The specific acceleration rate can be directly compared with the predictions of theoretical models of particle acceleration (see Sect. 1.7) and so provides a very useful constraint on the electron acceleration process [82, 83].

7.5 Hard X-Ray Imaging and the Global Energetics of Solar Flares

A series of papers [63, 64, 66] have dealt with the partition of the energy released in solar eruptive events[2] (flares and their associated coronal mass ejections). The released energy is manifested in a number of forms, including the acceleration of both electrons and ions (mostly protons but also heavier ions), the bulk kinetic energy of ejected plasma in the CME, and enhanced thermal energy in the confined flaring plasma. The thermal energy (ergs) in an ionized plasma with its two monatomic (electrons + protons) components, each with three degrees of freedom, is

$$E_{\rm thermal} = 2 \times \int \frac{3}{2} n(\mathbf{r})\, k_B\, T(\mathbf{r})\, dV \ , \tag{7.21}$$

where $n(\mathbf{r})$ (cm^{-3}) and $T(\mathbf{r})$ (K) are the number density and temperature at position $\mathbf{r}$ in the source, k_B (erg K^{-1}) is Boltzmann's constant, and the factor 2 comes from consideration of both electron and proton components. This may be approximated as

$$E_{\rm thermal} = 3\,\overline{n}\, k_B\, \overline{T}\, V \ , \tag{7.22}$$

[2] More recently, in a further series of papers summarized in [9], the results were presented of a comprehensive analysis of the global energetics of 399 solar M- and X-class flare events observed during the first 3.5 years of the Solar Dynamics Observatory (SDO) mission (2010–2013). The best effort was made to determine the magnetic, thermal, and nonthermal energies of each flare and the energies of the associated CMEs. The claim is made that this was "the first statistical study that establishes energy closure in solar flare-CME events," meaning that the sum of all the flare and CME energies is (within uncertainties) equal to the dissipated magnetic energy. The nonthermal energies in flare-accelerated electrons were revised in [10], with an emphasis on using a consistent values of the low energy cutoff (cf. Eq. (1.27)) to the electron spectra, but the main conclusion concerning closure was not changed. However, [199] pointed out that there are "contradicting results on energy partition obtained by various recent studies." These differences may be caused by changes in the thermal-nonthermal energy partition with flare strength. If this is the case, then the authors claim that "(a) an additional direct (i.e., non-beam) heating mechanism has to be present, and (b) considering that the bolometric emission originates mainly from deeper atmospheric layers, conduction or waves are required as additional energy transport mechanisms.

where $\overline{n}$ and $\overline{T}$ are the mean density and temperature in the region. To obtain estimates of $\overline{n}$ and $\overline{T}$ we use the fact that the free-free (bremsstrahlung) spectrum (photons cm^{-2} s^{-1} keV^{-1}) from a hot gas, observed at a distance R from the source, is

$$I(\epsilon) = \frac{C}{\epsilon\, R^2} \int \frac{n^2(\mathbf{r})}{T^{1/2}(\mathbf{r})} e^{-\epsilon/k_B T(\mathbf{r})}\, dV \simeq \frac{C}{\epsilon\, R^2\, \overline{T}^{1/2}}\, \mathrm{EM}\, e^{-\epsilon/k_B \overline{T}} \;, \tag{7.23}$$

where $C = 8.1 \times 10^{-39}\,\mathrm{cm}^3\ \mathrm{s}^{-1}\ \mathrm{K}^{1/2}$ [25, 45] and $R = 1.5 \times 10^{13}\,\mathrm{cm}$ is the Earth-Sun distance. Fitting the observed spectrum $I(\epsilon)$ with this expression provides estimates of the average source temperature $\overline{T}$ and the *emission measure* EM= $\int n^2(\mathbf{r})\, dV \simeq \overline{n}^2\, V$. Now, the expression for thermal energy (7.22) involves the quantity $3\,\overline{n}\, k_B\, \overline{T} V$, while the expression for the emission measure involves the quantity $\overline{n}^2\, V$, and so one cannot straightforwardly express the thermal energy E_{thermal} in terms of the fitted emission measure EM; the quantities $\overline{n}$ and V have to be determined independently. In practice this is done by rewriting equation (7.22) in the form

$$E_{\mathrm{thermal}} = 3\,\overline{n}\, k_B\, \overline{T}\, V = 3\, k_B\, \overline{T}\, \mathrm{EM}^{1/2}\, V^{1/2} \;, \tag{7.24}$$

which shows that to determine the thermal energy content from the spectral fit, the source volume V must also be estimated. The accuracy of the estimate for the energetic content of the thermal plasma therefore depends on the reliability of the estimate for V. In [66] and [63] the volume of the X-ray source was estimated by directly measuring the area A of the hard X-ray source on the plane of the sky and assuming an approximately isotropic source volume, so that $V \simeq A^{3/2}$; this leads to the thermal energy content E_{thermal} being $\propto A^{3/4}$, or $L^{3/2}$, where L is the characteristic dimension of the source. Combining the results of this analysis with measures of the energetic content of other features of the solar eruptive events studied (e.g., the pre-flare magnetic field, accelerated nonthermal particles, kinetic energy of the CME) allowed a more comprehensive understanding [66] to be developed of the production and flow of energy in an SEE.

Chapter 8
Future Possibilities

Abstract In this concluding chapter we present a brief synopsis of the future for hard X-ray imaging of solar flares, using both instruments that are currently operational and those proposed for future deployment.

The level of solar activity, including flares, exhibits a well-known cycle with a period of about 11 years. After a series of delays, *RHESSI* was launched in February 2002, toward the end of the peak in solar activity known as solar cycle 23 (Fig. 8.1). As was fully expected, over the course of many years of operation, the high-purity germanium detectors on *RHESSI*, and their Stirling-cycle mechanical cooler, gradually deteriorated, and several anneals were performed over the duration of the mission to partially restore the detectors to a healthy state. However, after 16 years of operation (many times the originally-projected 2-year lifetime of the mission), the operational status of the detectors had started to drop below the level where useful scientific measurements could be obtained. Finally, following a series of communications issues with the satellite that rendered science measurements impossible, the *RHESSI* instrument/satellite was decommissioned in October 2018; the satellite remains in Earth orbit but is expected to re-enter the atmosphere as early as 2022. During its period of science operations from 2002–2018, corresponding to the later phase of solar cycle 23 and most of solar cycle 24, *RHESSI* captured 115,000 flares at energies above 6 keV, including 468 at energies above 100 keV.

At the time of this writing, the Sun is emerging from the period of minimum activity at the end of solar cycle 24. An increase in flare rate should occur over the next few years, leading to a maximum in solar cycle 25 expected near the year 2025—see Fig. 8.1. However, the time taken to develop new instruments, test them in laboratories and on high-altitude balloons and/or suborbital rocket flights, obtain approval and funding for deployment on long-term orbital missions (e.g., SMEX or MidEX missions), and finally build and launch them takes many years. Hence only those missions that are already approved and funded can be expected to be operational during this upcoming solar maximum. These instruments, and their capabilities, are now described.

M. Piana et al., *Hard X-Ray Imaging of Solar Flares*,
https://doi.org/10.1007/978-3-030-87277-9_8

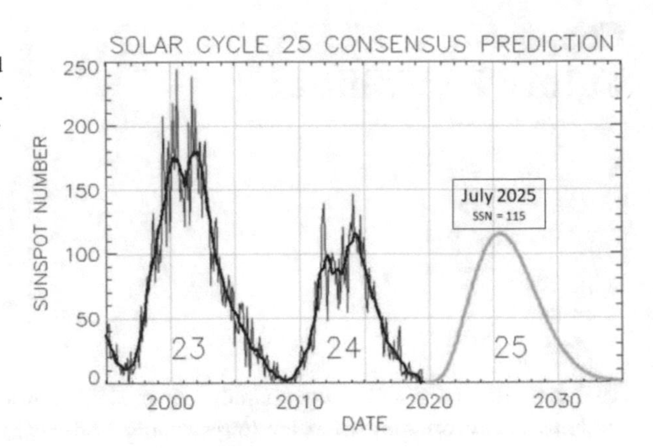

Fig. 8.1 Sunspot numbers for solar cycles 23 and 24 and predicted values for Cycle 25. The Plot is from https://www.swpc.noaa.gov/news/solar-cycle-25-forecast-update based on a forecast published on December 09, 2019, by a NOAA/NASA co-chaired international panel. The consensus was that the peak will be in July, 2025 (±8 months), with a smoothed sunspot number (SSN) of 115.

8.1 *STIX*

The primary hard X-ray imaging observations during the upcoming solar maximum will be made by the *STIX* instrument [120] already described in Sect. 3.6. *STIX* is on the ESA Solar Orbiter mission [72] launched on 2002 February 9. In the course of its first few years, the Solar Orbiter heliocentric orbit, perturbed by several planned planetary flybys, will cause it to pass as close as ∼0.28 AU to the Sun. At this distance, even modest *STIX* angular resolution down to 7″ results in a *spatial* resolution (in km) comparable to that of *RHESSI* in Earth orbit. Thus, *STIX* will provide spatio-spectral information on hard X-ray sources with similar resolution to that from *RHESSI*. It is intended to supplement these observations with observations from Earth orbit using instruments on CubeSats. One such instrument is MiSolFA [122]. Such observations would provide unprecedented *stereoscopic* observations that permit three-dimensional imaging of solar hard X-ray sources (cf. the CAT-scan medical imaging technique discussed in Sect. 2.1).

The Solar Orbiter nominal mission duration is 7 years with a possible additional 3 years of extended operations. *STIX* has already begun operations and has observed several small flares during the commissioning phase between 2020 June 5 and June 21 [16]. As discussed in Sect. 3.7, *STIX* is designed to provide spatial Fourier components of solar flares over a large range of flare sizes at energies between 4 and 150 keV and with angular resolutions from 7 to 180″. Importantly, the Solar Orbiter mission also contains a suite of instruments designed to measure both photons from the Sun and *in situ* particles and fields in interplanetary space. This information will allow the *STIX* observations to be placed in a more comprehensive context.

8.2 NuSTAR

Several solar flare observations have already been made with the Nuclear Spectroscopic Telescope ARsay (NuSTAR) [85]. Indeed, the first evidence of nonthermal hard X-ray emission from the Sun using a direct-focusing instrument covering energies from 4 to 12 keV, was reported in [78]. Occasional solar observations are expected to continue for the duration of this astrophysics mission, but they will have to be limited to periods of low solar activity since NuSTAR is so highly sensitive that it becomes saturated with high counting rates and consequent high deadtime on even non-flaring active regions and modest GOES A-class events.

8.3 *FOXSI*

Given the tremendous success of the astrophysics NuSTAR mission [85], that uses focusing optics to provide very high sensitivity up to energies of 79 keV, it seems inevitable that a similar instrument optimized for solar flare observations will eventually be flown. To that end, the Focusing Optics X-ray Solar Imager (FOXSI) was proposed as part of the Solar Eruptive Event 2020 Mission Concept white paper submitted for the 2013 Heliophysics Decadal Survey [126]. A version of FOXSI has been developed and flown on several suborbital rocket flights [11, 197]. Excellent images of the soft X-ray emission up to $\sim$10 keV have already been made on previous rocket flights and additional flights of the FOXSI instrument previously flown are planned with various improvements to the optics and detectors [79].

Unfortunately, neither the proposed SMEX version of FOXSI [39], nor the MidEx version, the Fundamentals of Impulsive Energy Release in the Corona Explorer (FIERCE) that included FOXSI [172], were selected for flight by NASA. Consequently, it will probably not be possible to make the first hard X-ray images from a satellite with focusing optics optimized for solar observations during the upcoming maximum of solar cycle 25.

Although FOXSI is capable of imaging up to $\sim$80 keV, the chances are negligible for detecting a flare with emission up to this energy during a rocket flight launched at a random time since the observing period lasts typically for only about 6 min. However, an exciting plan to increase the chances of catching a flare is to include FOXSI in NASA's first solar flare sounding rocket campaign[1] [165]. The plan is to ready the rocket and payload for flight but wait for a flare on the launch pad at the Poker Flat Research Range in Alaska, possibly for weeks. Two such rockets with different payloads will be launched near-simultaneously within minutes of a signal that a suitable flare is in progress. The probability of catching a C1 flare with this scheme is estimated to be $\geq$95% in 2023–2026 and $\geq$30% for an M1 flare.

[1] https://rscience.gsfc.nasa.gov/keydocs/SolarFlareCampaign.pdf.

There are currently no plans to obtain flare observations by flying FOXSI on a long-duration high-altitude balloon, similar to the flight of an early version on the High Energy Replicated Optics to Explore the Sun (HEROES) mission in 2013 [38]. Such a flight would allow observations above the $\sim$20 keV cutoff of the overlying atmosphere, either for the several hours of a typical flight or for several days or even weeks for a long-duration flight from the Antarctic or other location.

8.4 Advanced Spaced-Based Solar Observatory

The Chinese are planning [207] to launch HXI on ASO-S in 2022. It will measure spatial Fourier components using about 92 bi-grid collimators and corresponding Lanthanum Bromide (LaBr) detectors. Angular resolutions as fine as $6''$ will be achieved at energies between 30 and 200 keV.

8.5 GRIPS

The Gamma-Ray Imager/Polarimeter for Solar flares (GRIPS) was part of the Solar Eruptive Event 2020 (SEE 2020) Mission Concept white paper submitted for the 2010 Heliophysics Decadal Survey [126]. The science objective of this instrument is to obtain information on the protons and heavier ions that are accelerated along with the electrons that produce the bremsstrahlung X-rays during flares. The basic concept was to achieve high-resolution gamma-ray imaging in the 100 keV to $>$15 MeV energy range with a factor of $\gtrsim$10 increase in sensitivity, and better than $10''$ FWHM resolution compared to *RHESSI*'s $35''$ resolution at 2.2 MeV. GRIPS was flown [55] on a high altitude Antarctic balloon flight in January 2016, and has been funded for a second flight during the upcoming increase in solar activity in solar cycle 25.

8.6 Hard X-ray Polarimetry

The full bremsstrahlung cross-section (Eq. (1.8)) is a function not only of the energies of the exciting electron and emitted photon; it also depends on the direction of the emitted photon relative to the plane in which the bremsstrahlung electron-ion collision occurs. There are different cross-sections for photon polarization in, and perpendicular to, that plane—see [105] and the Appendix in [132]. If the exciting electron distribution is isotropic, then, by symmetry of the various collision planes involved, the total emitted bremsstrahlung will be unpolarized. However, if the electron distribution is anisotropic (e.g., beamed in the direction of the guiding magnetic field due to acceleration by field-aligned electric fields), then the emitted bremsstrahlung will have a net polarization [24], with the degree of polarization depending on the degree of anisotropy in the electron distribution and the direction of the polarization vector determined by the direction in which the electron flux is concentrated.

Measuring the polarization of hard X-rays from solar flares thus offers important unique diagnostic potential for understanding the flare-accelerated electrons (e.g., [27, 86, 100, 123]). Indeed, with the location, spectrum, and intensity of solar flare hard X-ray radiation now well-characterized, the polarization state of the emitted photon field remains the only quantity yet to be reliably measured. Even thermal hard X-ray emission can result [60] in nonzero polarization due to anisotropy in the underlying electron phase-space distribution function $f(\mathbf{v})$ in the presence of strong temperature gradients. Nevertheless, the detection of large X-ray polarization would "be strong support for the nonthermal model" [123]. The ubiquitous *RHESSI* observations of chromospheric hard X-ray footpoints, especially at higher energies (e.g., the 2002 February 20 and 2002 July 23 events studied in Chap. 7), suggests that significant levels of polarization should indeed be present, at least at higher photon energies. Such measurements would add valuable insight into the processes of acceleration and transport of electrons in the flaring atmosphere.

A recent paper [101] discusses this in some detail, specifically in the context of prospective X-ray polarimetry missions. There it is argued that hard X-ray spectro-polarimetry, even without imaging, can be a very valuable tool in constraining both the anisotropy and the spectrum (in particular at high energies) of the accelerated electron distribution. Clearly, however, measuring the degree of polarization as a function of position on the Sun with arcsecond resolution is a necessary step towards the ultimate goal of understanding the nature of the emitting electrons.

Despite this recognition of the immense value of hard X-ray polarization information, no reliable measures of such have been achieved to date (see discussion in [123]). An attempt was made with *RHESSI* to measure the spatially integrated linear polarization in 20–100 keV X-rays by simply adding a small (passive) beryllium scattering cylinder, strategically located [135] between the germanium detectors in such a way that photons that were Compton-scattered from it could be detected in the rear segments of the detectors. Deka-keV photons that pass through the imaging grids are stopped in the detector front segment and cannot reach the rear segment. Thus, solar deka-keV photons that are scattered in the beryllium cylinder may be detected through the side of the rear segments. If the incoming flux is polarized, then the count rates in the various detector rear segments that can "see" the scatterer will be modulated at half the spacecraft rotation period ($\sim$2 s), with the modulation amplitude proportional to the degree of polarization of the bremsstrahlung flux and the phase depending on the direction of the polarization vector on the plane of the sky.[2] Unfortunately, systematic effects introduced by other

[2] The reference plane containing both the line-of-sight to the source and the local solar vertical corresponds to the radial direction on the solar disk. A radial [transverse] orientation of the polarization vector therefore corresponds to a preponderance of emission parallel to [perpendicular to] this reference plane. The orientation of the polarization vector thus provides [24] information on

sources of count-rate modulations in the rear segments proved to be so large that it was not possible to definitively identify any X-ray polarization, even under the most favorable circumstances for the largest flares detected [111, 136].

The astrophysical importance of hard X-ray polarization for understanding many different sources has generated strong programs to develop the capability to make the required measurements (e.g., [35]). The hard X-rays have polarization signatures that provide unique information that is not available from current spectroscopic and timing observations of a wide variety of objects including black holes, active galactic nuclei, neutron starts, gamma-ray burst, magnetars, blazars, pulsars, etc. Several instruments designed specifically to measure hard X-ray polarization are in various stages of planning or in preparation for launch. The Imaging X-ray Polarimetry Explorer (IXPE) [201] will be launched in the fall of 2021. It will measure astrophysical sources in the soft X-ray energy range from 2–8 keV with an angular resolution of $<30''$ HPD.

Plans for making solar flare polarization measurements are not so far advanced as in astrophysics. Three hard X-ray polarimeters specifically optimized for the intense fluxes in solar flares are discussed in [101]. The first of these is GRIPS, discussed in Sect. 8.5 above, that will perform imaging spectro-polarimetry of solar flares in the 150 keV to 10 MeV range, with an angular resolution as fine as $12''.5$. A second is The SolAr Polarimeter for Hard x-Rays (SAPPHIRE) [163], a single module with a CubeSat form factor designed to measure hard X-ray polarization from $\sim$5 to $\sim$100 keV. The Solar HARd x-ray Polarimer/Imager Experiment (SHARPIE) [163] is a concept that would array several SAPPHIRE modules behind bi-grid subcollimators to produce images of the polarized flux, with angular resolutions similar to that achieved with *RHESSI*. A third polarimeter is on the Japanese PhoENiX[3] (Physics of Energetic and Non-thermal plasmas in the X (= magnetic reconnection) region) mission [141, 142] planned for a 2030 launch. This will cover the energy range from 20 to 600 keV but with no imaging capability.

The Chinese gamma-ray burst instrument POLAR [209] on board the TianGong2 space lab, launched on 2016 September 15, is designed to measure the linear polarization of hard X-rays in the 50–500 keV energy range.

Finally, a novel hard X-ray imaging spectropolarimeter that utilizes the combination of a fine-pixel complementary metal-oxide semiconductor (CMOS) imaging sensor and a narrow field-of-view coded-aperture mask with multiple different random patterns to achieve $30''$ angular resolution in the 10–30 keV energy range has also been proposed [147]. While these characteristics will not be adequate to separate the different source features even in a typical large flare, they can demonstrate the capability of a concept that could be extended for future larger missions using an X-ray collection mirror.

---

the anisotropy of the electron distribution with respect to the solar vertical, corresponding roughly to the direction of the guiding magnetic field.

[3] The symbol *X* denotes the type of magnetic field geometry commonly associated with the magnetic reconnection region.

8.7 Conclusion

In its sixteen-plus years of scientific operation, the *RHESSI* instrument, with its pioneering combination of high spectral resolution detectors and rotating modulation collimator imaging capability, revolutionized our understanding of how high-energy particles are accelerated and propagate during solar flares and eruptive events. The success of *RHESSI* has generated a series of follow-on missions aimed at building on its legacy as outlined in this chapter.

While a great deal of *RHESSI*'s success clearly derived from its unique instrumental capabilities, equally important was the development of innovative algorithms to extract the maximum scientific information from the native data set: a noisy set of sparsely-distributed spatial Fourier components (Sect. 3.2). This book has summarized these methods, their strengths and limitations, and the ways in which they have been applied to *RHESSI* data in order to enhance our understanding of solar flares. As noted in Sect. 3.3, the (very deliberate) way in which the *RHESSI* data were transmitted and stored (with each detected photon time- and energy-tagged) allows the original data to be analyzed using potentially even more sophisticated methods yet to be developed. With this capability in mind, at the time of this writing efforts are underway to establish an archive of *RHESSI* data for use by researchers in both the near- and far-future. It will reside at, and be freely available on-line from, the Solar Data Analysis Center (SDAC) at Goddard Space Flight Center.[4]

High-energy imaging of solar flares thus remains an active field, with many intriguing prospects. And, since some of the most exciting science lies at the boundaries of disciplines, the collaboration formed between solar scientists, instrument developers, mathematicians, and developers of computational algorithms that was so essential to *RHESSI*'s success, has opened up the possibility for many more, similarly fruitful, investigations into the future.

[4] https://umbra.nascom.nasa.gov/index.html/.

References

1. L.W. Acton, J.L. Culhane, A.H. Gabriel, R.D. Bentley, J.A. Bowles, J.G. Firth, M.L. Finch, C.W. Gilbreth, P. Guttridge, R.W. Hayes, E.G. Joki, B.B. Jones, B.J. Kent, J.W. Leibacher, R.A. Nobles, T.J. Patrick, K.J.H. Phillips, C.G. Rapley, P.H. Sheather, J.C. Sherman, J.P. Stark, L.A. Springer, R.F. Turner, and C.J. Wolfson, The Soft X-ray polychromator for the solar maximum mission. Solar Phys. **65**(1), 53–71 (1980)
2. M. Alaoui, G.D. Holman, Understanding breaks in flare X-ray spectra: evaluation of a cospatial collisional return-current model. Astrophys. J., **851**(2), 78 (2017)
3. M. Alaoui, G.D. Holman, J.C. Allred, R.T. Eufrasio. Role of suprathermal runaway electrons returning to the acceleration region in solar flares. Astrophys. J. **917**(2), 74 (2021)
4. S.K. Antiochos, P.A. Sturrock. Evaporative cooling of flare plasma. Astrophys. J. **220**, 1137–1143 (1978)
5. M.J. Aschwanden, *The Sun* (Elsevier, Amsterdam, 2007), pp. 71–98
6. M.J. Aschwanden, J.C. Brown, E.P. Kontar, Chromospheric height and density measurements in a solar flare observed with RHESSI II. Data analysis. Solar Phys. **210**, 383–405 (2002)
7. M.J. Aschwanden, E. Schmahl, The RHESSI Team, Reconstruction of RHESSI solar flare images with a forward fitting method. Solar Phys. **210**, 193–211 (2002)
8. M.J. Aschwanden, T.R. Metcalf, S. Krucker, J. Sato, A.J. Conway, G.J. Hurford, E.J. Schmahl, On the photometric accuracy of RHESSI imaging and spectrosocopy. Solar Phys. **219**, 149–157 (2004)
9. M.J. Aschwanden, A. Caspi, C.M.S. Cohen, G. Holman, J. Jing, M. Kretzschmar, E.P. Kontar, J.M. McTiernan, R.A. Mewaldt, A. O'Flannagain, I.G. Richardson, D. Ryan, H.P. Warren, Y. Xu, Global energetics of solar flares. V. Energy closure in flares and coronal mass ejections. Astrophys. J. **836**(1), 17 (2017)
10. M.J. Aschwanden, E.P. Kontar, N.L.S. Jeffrey. Global energetics of solar flares. VIII. The low-energy cutoff. Astrophys. J. **881**(1), 1 (2019)
11. P.S. Athiray, J. Vievering, L. Glesener, S. Ishikawa, N. Narukage, J.C. Buitrago-Casas, S. Musset, A. Inglis, S. Christe, S. Krucker, D. Ryan, FOXSI-2 solar microflares. I. Multi-instrument differential emission measure analysis and thermal energies. Astrophys. J. **891**(1), 78 (2020)
12. M. Avriel, *Nonlinear Programming: Analysis and Methods* (Courier Corporation, North Chelmsford, 2003)
13. C. Badenes, X-ray studies of supernova remnants: a different view of supernova explosions. Proc. Nat. Acad. Sci. **107**(16), 7141–7146 (2010)

M. Piana et al., *Hard X-Ray Imaging of Solar Flares*,
https://doi.org/10.1007/978-3-030-87277-9

14. D.N. Baker, The major solar eruptive event in July 2012: Defining extreme space weather scenarios (Invited), in *AGU Fall Meeting Abstracts*, vol. 2013 (2013), pp. SM13C–04
15. M. Battaglia, E.P. Kontar, I.G. Hannah, The influence of albedo on the size of hard X-ray flare sources. Astron. Astrophys. **526**, A3+ (2011)
16. A.F. Battaglia, J. Saqri, P. Massa, E. Perracchione, E.C.M. Dickson, H. Xiao, A.M. Veronig, A. Warmuth, M. Battaglia, G.J. Hurford, A. Meuris, O. Limousin, L. Etesi, S.A. Maloney, R.A. Schwartz, M. Kuhar, F. Schuller, V. Senthamizh Pavai, S. Musset, D.F. Ryan, L. Kleint, M. Piana, A.M. Massone, F. Benvenuto, J. Sylwester, M. Litwicka, M. Stęślicki, T. Mrozek, N. Vilmer, F. Fárník, J. Kašparová, G. Mann, P.T. Gallagher, B.R. Dennis, A. Csillaghy, A.O. Benz, S. Krucker. STIX X-ray microflare observations during the solar orbiter commissioning phase (2021). arXiv:2106.10058
17. T. Bayes, R. Price, An Essay towards solving a problem in the doctrine of chance. Philos. Trans. Roy. Soc. Lond. **53**, 370 (1763)
18. A. Beck, M. Teboulle, Fast gradient-based algorithms for constrained total variation image denoising and deblurring problems. IEEE Trans. Image Process. **18**, 2419 (2011)
19. F. Benvenuto, R. Schwartz, M. Piana, A.M. Massone, Expectation maximization for hard X-ray count modulation profiles. Astron. Astrophys. **555**, A61 (2013)
20. F. Benvenuto, H. Haddar, B. Lantz, A robust inversion method according to a new notion of regularization for poisson data with an application to nanoparticle volume determination. SIAM J. Appl. Math. **76**(1), 276–292 (2016)
21. R.L. Blake, T.A. Chubb, H. Friedman, A.E. Unzicker. Interpretation of X-ray photograph of the sun. Astrophys. J. **137**, 3 (1963)
22. S.-C. Bong, J. Lee, D.E. Gary, H.S. Yun. Spatio-spectral maximum entropy method. I. Formulation and test. Astrophys. J. **636**, 1159–1165 (2006)
23. J.C. Brown, The deduction of energy spectra of non-thermal electrons in flares from the observed dynamic spectra of hard X-ray bursts. Solar Phys. **18**, 489–502 (1971)
24. J.C. Brown, The directivity and polarisation of thick target X-ray bremsstrahlung from solar flares. Solar Phys. **26**, 441 (1972)
25. J.C. Brown, On the thermal interpretation of hard X-ray bursts from solar flares, in *Coronal Disturbances, IAU Symposium*, ed. by G.A. Newkirk, vol. 57 (1974), p. 395
26. J.C. Brown, A.L. MacKinnon, Bremsstrahlung spectra from thick-target electron beams with noncollisional energy losses. Astrophys. J. Lett. **292**, L31–L34 (1985)
27. J.C. Brown, A.N. McClymont, I.S. McLean, Interpretation of solar hard X-ray burst polarization measurements. Nature **247**, 448–449 (1974)
28. J.C. Brown, M.J. Aschwanden, E.P. Kontar. Chromospheric height and density measurements in a solar flare observed with RHESSI I. Theory. Solar Phys., **210**, 373–381 (2002)
29. J.C. Brown, A.G. Emslie, E.P. Kontar, The determination and use of mean electron flux spectra in solar flares. Astrophys. J. Lett. **595**, L115–L117 (2003)
30. J.C. Brown, A.G. Emslie, G.D. Holman, C.M. Johns-Krull, E.P. Kontar, R.P. Lin, A.M. Massone, M. Piana. Evaluation of algorithms for reconstructing electron spectra from their bremsstrahlung hard X-ray spectra. Astrophys. J. **643**, 523–531 (2006)
31. J.C. Brown, J. Kašparová, A.M. Massone, M. Piana, Fast spectral fitting of hard X-ray bremsstrahlung from truncated power-law electron spectra. Astron. Astrophys. **486**(3), 1023–1029 (2008)
32. H. Carmichael, A Process for flares. NASA Spec. Publ. **50**, 451 (1964)
33. R.C. Carrington, Description of a singular appearance seen in the sun on September 1, 1859. Monthly Notices Roy. Astron. Soc. **20**, 13–15 (1859)
34. W. Cash, Parameter estimation in astronomy through application of the likelihood ratio. Astrophys. J. **228**, 939–947 (1979)
35. T. Chattopadhyay, Hard X-ray polarimetry: An overview of the method, science drivers and recent findings (2021). arXiv:2104.05244
36. F.F. Chen, *Introduction to Plasma Physics and Controlled Fusion*. (Springer, Berlin, 2016)
37. B. Chen, T.S. Bastian, The role of inverse compton scattering in solar coronal hard X-ray and γ-ray sources. Astrophys. J. **750**(1), 35 (2012)

38. S.D. Christe, A. Shih, M. Rodriguez, A. Cramer, K. Gregory, M. Edgerton, J. Gaskin, C. Wilson-Hodge, J. Apple, K. Stevenson Chavis, A. Jackson, L. Smith, K. Dietz, B. O'Connor, A. Sobey, H. Koehler, B. Ramsey, The high energy replicated optics to explore the sun mission: a hard X-ray balloon-borne telescope, in *Solar Physics and Space Weather Instrumentation V* , ed. by S. Fineschi, J. Fennelly. Society of Photo-Optical Instrumentation Engineers (SPIE) Conference Series vol. 8862 (2013), p. 886206
39. S.D. Christe, A.Y. Shih, B.R. Dennis, L. Glesener, S. Krucker, P. Saint-Hilaire, M. Gubarev, B. Ramsey, The focusing optics X-ray solar imager small explorer concept mission, in *AAS/Solar Physics Division Abstracts #47*. AAS/Solar Physics Division Meeting, vol. 47 (2016), p. 8.02
40. M.J. Cieślak, K.A.A. Gamage, R. Glover, Coded-aperture imaging systems: past, present and future development - a review. Radiation Measur. **92**, 59–71 (2016)
41. M. Clerc, *Particle Swarm Optimization*, vol. 93 (Wiley, Hoboken, 2010)
42. A. Codispoti, G. Torre, M. Piana, N. Pinamonti, Return currents and energy transport in the solar flaring atmosphere. Astrophys. J. **773**, 121 (2013)
43. P.L. Combettes, J.-C. Pesquet, Proximal splitting methods in signal processing, in *Fixed-Point Algorithms for Inverse Problems in Science and Engineering*, ed. by H. Bauschke, R. Burachik, P. Combettes, V. Elser, D. Luke, H. Wolkowicz (2011), pp. 185–212
44. I.J.D. Craig, J.C. Brown, *Inverse Problems in Astronomy: a Guide to Inversion Strategies for Remotely Sensed Data* (Adam Hilger, Bristol, 1986)
45. J.L. Culhane, Thermal continuum radiation from coronal plasmas at soft X-ray wavelengths. Monthly Notices Roy. Astron. Soc. **144**, 375–+ (1969)
46. D.W. Datlowe, Errata: pulse pile-up in hard X-ray detector systems [Space Sci. Instrum., vol. 1, p. 389–406 (1975)]. Space Sci. Instrum. **2**, 523–+ (1976)
47. D.W. Datlowe, Summary of "Pulse Pile-up in Solar Hard X-Ray Detector Systems". Space Sci. Instrum. **2**, 239–+ (1976)
48. D.W. Datlowe, Pulse pile-up in nuclear particle detection systems with rapidly varying counting rates. Nuclear Instrum. Methods **145**(2), 379–387 (1977)
49. B.R. Dennis, R.L. Pernak, Hard X-ray flare source sizes measured with the ramaty high energy solar spectroscopic imager. Astrophys. J. **698**, 2131–2143 (2009)
50. B.R. Dennis, A.K. Tolbert, A remarkably narrow RHESSI X-ray flare on 2011 September 25. Astrophys. J. **887**(2), 131 (2019)
51. B.R. Dennis, G.K. Skinner, M.J. Li, A.Y. Shih, Very high-resolution solar X-ray imaging using diffractive optics. Solar Phys. **279**(2), 573–588 (2012)
52. B.R. Dennis, M.A. Duval-Poo, M. Piana, A.R. Inglis, A.G. Emslie, J. Guo, Y. Xu, Coronal hard X-ray sources revisited. Astrophys. J. **867**(1), 82 (2018)
53. R.H. Dicke, Scatter-hole cameras for X-rays and gamma rays. Astrophys. J. Lett. **153**, L101 (1968)
54. H. Dreicer, Electron and Ion runaway in a fully ionized gas. I. Phys. Rev. **115**(2), 238–249 (1959)
55. N. Duncan, P. Saint-Hilaire, A.Y. Shih, G.J. Hurford, H.M. Bain, M. Amman, B.A. Mochizuki, J. Hoberman, J. Olson, B.A. Maruca, N.M. Godbole, D.M. Smith, J. Sample, N.A. Kelley, A. Zoglauer, A. Caspi, P. Kaufmann, S. Boggs, R.P. Lin. First flight of the Gamma-Ray imager/polarimeter for solar flares (GRIPS) instrument, in *Space Telescopes and Instrumentation 2016: Ultraviolet to Gamma Ray*, ed. by J.-W.A. den Herder, T. Takahashi, M. Bautz. Society of Photo-Optical Instrumentation Engineers (SPIE) Conference Series, vol. 9905 (2016), p. 99052Q
56. P. Durouchoux, H. Hudson, J. Matteson, G. Hurford, K. Hurley, E. Orsal, Gamma-ray Imaging with a rotating modulator. Astron. Astrophys. **120**(1), 150–155 (1983)
57. M.A. Duval-Poo, M. Piana, A.M. Massone, Solar hard X-ray imaging by means of compressed sensing and finite isotropic wavelet transform. Astron. Astrophys. **615**, A59 (2018)
58. A.G. Emslie, The collisional interaction of a beam of charged particles with a hydrogen target of arbitrary ionization level. Astrophys. J. **224**, 241–246 (1978)

59. A.G. Emslie, The effect of reverse currents on the dynamics of nonthermal electron beams in solar flares and on their emitted X-ray bremsstrahlung. Astrophys. J. **235**, 1055–1065 (1980)
60. A.G. Emslie, J.C. Brown, The polarization and directivity of solar-flare hard X-ray bremsstrahlung from a thermal source. Astrophys. J. **237**, 1015–1023 (1980)
61. A.G. Emslie, R.K. Barrett, J.C. Brown, An empirical method to determine electron energy modification rates from spatially resolved hard X-ray data. Astrophys. J. **557**, 921–929 (2001)
62. A.G. Emslie, E.P. Kontar, S. Krucker, R.P. Lin, RHESSI hard X-ray imaging spectroscopy of the large Gamma-Ray flare of 2002 July 23. Astrophys. J. Lett. **595**, L107–L110 (2003)
63. A.G. Emslie, H. Kucharek, B.R. Dennis, N. Gopalswamy, G.D. Holman, G.H. Share, A. Vourlidas, T.G. Forbes, P.T. Gallagher, G.M. Mason, T.R. Metcalf, R.A. Mewaldt, R.J. Murphy, R.A. Schwartz, T.H. Zurbuchen, Energy partition in two solar flare/CME events. J. Geophys. Res. (Space Phys.) **109**(A18), 10104 (2004)
64. A.G. Emslie, B.R. Dennis, G.D. Holman, H.S. Hudson. Refinements to flare energy estimates: a followup to "Energy Partition in Two Solar Flare/CME Events" by A. G. Emslie et al. J. Geophys. Res. (Space Phys.) **110**(A9), 11103 (2005)
65. A.G. Emslie, G.J. Hurford, E.P. Kontar, A.M. Massone, M. Piana, M. Prato, Y. Xu, Determining the spatial variation of accelerated electron spectra in solar flares, in *American Institute of Physics Conference Series*, ed. by G. Li, Q. Hu, O. Verkhoglyadova, G.P. Zank, R.P. Lin, J. Luhmann. American Institute of Physics Conference Series, vol. 1039 (2008), pp. 3–10
66. A.G. Emslie, B.R. Dennis, A.Y. Shih, P.C. Chamberlin, R.A. Mewaldt, C.S. Moore, G.H. Share, A. Vourlidas, B.T. Welsch, Global energetics of thirty-eight large solar eruptive events. Astrophys. J. **759**, 71 (2012)
67. H.W. Engl, C.W. Groetsch, *Inverse and Ill-Posed Problems*, vol. 4 (Elsevier, Amsterdam, 2014)
68. S. Enome, HINOTORI - a Japanese satellite for solar flare studies. Adv. Space Res. **2**(11), 201–202 (1982)
69. S. Felix, R. Bolzern, M. Battaglia, A compressed sensing-based image reconstruction algorithm for solar flare X-ray observations. Astrophys. J. **849**(1), 10 (2017)
70. W.H. Follett, L.T. Ostwald, J.O. Simpson, T.M. Spencer, Decade of improvements to orbiting solar observatories. J. Spacecraft Rockets **11**, 327 (1974)
71. C. Forbes, M. Evans, N. Hastings, B. Peacock, *Statistical Distributions*, chapter 35 (Wiley, Hoboken, 2010), pp. 152–156
72. T. Forveille, S. Shore, The solar orbiter mission. Astron. Astrophys., **642**, E1 (2020)
73. L. Foschini, What we talk about when we talk about blazars? Front. Astron. Space Sci. **4**, 6 (2017)
74. S.L. Freeland, B.N. Handy, Data analysis with the SolarSoft system. Solar Phys. **182**(2), 497–500 (1998)
75. W.-Q. Gan, C. Zhu, Y.-Y. Deng, H. Li, Y. Su, H.-Y. Zhang, B. Chen, Z. Zhang, J. Wu, L. Deng, Y. Huang, J.-F. Yang, J.-J. Cui, J. Chang, C. Wang, J. Wu, Z.-S. Yin, W. Chen, C. Fang, Y.-H. Yan, J. Lin, W.-M. Xiong, B. Chen, H.-C. Bao, C.-X. Cao, Y.-P. Bai, T. Wang, B.-L. Chen, X.-Y. Li, Y. Zhang, L. Feng, J.-T. Su, Y. Li, W. Chen, Y.-P. Li, Y.-N. Su, H.-Y. Wu, M. Gu, L. Huang, X.-J. Tang, Advanced space-based solar observatory (ASO-S):an overview. Res. Astron. Astrophys. **19**(11), 156 (2019)
76. N. Gehrels, G. Chincarini, P. Giommi, K.O. Mason, J.A. Nousek, A.A. Wells, N.E. White, S.D. Barthelmy, D.N. Burrows, L.R. Cominsky, K.C. Hurley, F.E. Marshall, P. Mészáros, P.W.A. Roming, L. Angelini, L.M. Barbier, T. Belloni, S. Campana, P.A. Caraveo, M.M. Chester, O. Citterio, T.L. Cline, M.S. Cropper, J.R. Cummings, A.J. Dean, E.D. Feigelson, E.E. Fenimore, D.A. Frail, A.S. Fruchter, G.P. Garmire, K. Gendreau, G. Ghisellini, J. Greiner, J.E. Hill, S.D. Hunsberger, H.A. Krimm, S.R. Kulkarni, P. Kumar, F. Lebrun, N.M. Lloyd-Ronning, C.B. Markwardt, B.J. Mattson, R.F. Mushotzky, J.P. Norris, J. Osborne, B. Paczynski, D.M. Palmer, H.S. Park, A.M. Parsons, J. Paul, M.J. Rees, C.S. Reynolds, J.E. Rhoads, T.P. Sasseen, B.E. Schaefer, A.T. Short, A.P. Smale, I.A. Smith, L. Stella, G. Tagliaferri, T. Takahashi, M. Tashiro, L.K. Townsley, J. Tueller, M.J.L. Turner, M. Vietri,

W. Voges, M.J. Ward, R. Willingale, F.M. Zerbi, W.W. Zhang, The Swift Gamma-Ray Burst Mission. Astrophys. J. **611**(2), 1005–1020 (2004)
77. S. Giordano, N. Pinamonti, M. Piana, A.M. Massone, The process of data formation for the spectrometer/telescope for imaging X-rays (STIX) in solar orbiter. SIAM J. Imag. Sci. **8**(2), 1315–1331 (2015)
78. L. Glesener, S. Krucker, J. Duncan, I.G. Hannah, B.W. Grefenstette, B. Chen, D.M. Smith, S.M. White, H. Hudson, Accelerated electrons observed down to <7 keV in a NuSTAR solar microflare. Astrophys. J. Lett. **891**(2), L34 (2020)
79. L. Glesener, J. Buitrago-Casas, J. Duncan, S. Nagasawa, A. Pantazides, S. Perez-Piel, Y. Zhang, J. Vievering, S. Musset, S. Panchapakesan, W. Baumgartner, S. Bongiorno, P. Champey, S. Christe, S. Courtade, H. Kanniainen, S. Krucker, S. Ishikawa, J. Martinez Oliveros, I. Mitsuishi, N. Narukage, E. Peretz, D. Ryan, T. Takahashi, S. Watanabe, A. Winebarger, High resolution FOXSI: The development Of FOXSI-4, in *American Astronomical Society Meeting Abstracts*. American Astronomical Society Meeting Abstracts, vol. 53 (2021), p. 313.01
80. G.H. Golub, M. Heath, G. Wahba, Generalized cross-validation as a method for choosing a good ridge parameter. Technometrics **21**(2), 215–223 (1979)
81. J. Guo, S. Liu, L. Fletcher, E.P. Kontar, Relationship between hard and soft X-ray emission components of a solar flare. Astrophys. J. **728**(1), 4 (2011)
82. J. Guo, A.G. Emslie, A.M. Massone, M. Piana, Properties of the acceleration regions in several loop-structured solar flares. Astrophys. J. **755**, 32 (2012)
83. J. Guo, A.G. Emslie, M. Piana, The specific acceleration rate in loop-structured solar flares — implications for electron acceleration models. Astrophys. J. **766**, 28 (2013)
84. P.C. Hansen, D.P. O'Leary, The use of the L-curve in the regularization of discrete Ill-posed problems. SIAM J. Sci. Comput. **14**(6), 1487–1503 (1993)
85. F.A. Harrison, W.W. Craig, F.E. Christensen, C.J. Hailey, W.W. Zhang, S.E. Boggs, D. Stern, W.R. Cook, K. Forster, P. Giommi, B.W. Grefenstette, Y. Kim, T. Kitaguchi, J.E. Koglin, K.K. Madsen, P.H. Mao, H. Miyasaka, K. Mori, M. Perri, M.J. Pivovaroff, S. Puccetti, V.R. Rana, N.J. Westergaard, J. Willis, A. Zoglauer, H. An, M. Bachetti, N.M. Barrière, E C. Bellm, V. Bhalerao, N.F. Brejnholt, F. Fuerst, C.C. Liebe, C.B. Markwardt, M. Nynka, J.K. Vogel, D.J. Walton, D.R. Wik, D.M. Alexander, L.R. Cominsky, A.E. Hornschemeier, A. Hornstrup, V.M. Kaspi, G.M. Madejski, G. Matt, S. Molendi, D.M. Smith, J.A. Tomsick, M. Ajello, D.R. Ballantyne, M. Baloković, D. Barret, F.E. Bauer, R.D. Blandford, W. N Brandt, L.W. Brenneman, J. Chiang, D. Chakrabarty, J. Chenevez, A. Comastri, F. Dufour, M. Elvis, A.C. Fabian, D. Farrah, C.L. Fryer, E.V. Gotthelf, J.E. Grindlay, D.J. Helfand, R. Krivonos, D.L. Meier, J.M. Miller, L. Natalucci, P. Ogle, E.O. Ofek, A. Ptak, S.P. Reynolds, J.R. Rigby, G. Tagliaferri, S.E. Thorsett, E. Treister, C.M. Urry, The nuclear spectroscopic telescope array (NuSTAR) high-energy X-ray mission. Astrophys. J. **770**(2), 103 (2013)
86. E. Haug, Polarization of hard X-rays from solar flares. Solar Phys. **25**, 425–+ (1972)
87. E. Haug, Proton-electron bremsstrahlung. Astron. Astrophys. **406**, 31–35 (2003)
88. T. Hirayama, Theoretical model of flares and prominences. I: Evaporating flare model. Solar Phys. **34**, 323–338 (1974)
89. R. Hodgson, On a Curious appearance seen in the sun. Monthly Notices Roy. Astron. Soc. **20**, 15–16 (1859)
90. J.A. Högbom, Aperture synthesis with a non-regular distribution of interferometer baselines. Astron. Astrophys. Suppl. **15**, 417 (1974)
91. G.D. Holman, Solar eruptive events. Phys. Today **65**, 4:56–57 (2012)
92. P. Hoyng, A. Duijveman, A. Boelee, C. de Jager, M. Galama, R. Hoekstra, J. Imhof, H. Lafleur, M.E. Machado, R. Fryer, Hard X-ray imaging of two flares in active region 2372. Astrophys. J. Lett. **244**, L153–L156 (1981)
93. P. Hoyng, A. Duijveman, M.E. Machado, D.M. Rust, Z. Svestka, A. Boelee, C. de Jager, K.T. Frost, H. Lafleur, G.M. Simnett, H.F. van Beek, B.E. Woodgate, Origin and location of the hard X-ray emission in a two-ribbon flare. Astrophys. J. Lett. **246**, L155+ (1981)

94. G.J. Hurford, X-ray imaging with collimators, masks and grids. ISSI Sci. Rep. Ser. **9**, 223–234 (2010)
95. G.J. Hurford, D.W. Curtis, The PMTRAS roll aspect system on RHESSI. Solar Phys. **210**, 101–113 (2002)
96. G.J. Hurford, E.J. Schmahl, R.A. Schwartz, A.J. Conway, M.J. Aschwanden, A. Csillaghy, B.R. Dennis, C. Johns-Krull, S. Krucker, R.P. Lin, J. McTiernan, T.R. Metcalf, J. Sato, D.M. Smith, The RHESSI imaging concept. Solar Phys. **210**, 61–86 (2002)
97. G.J. Hurford, E.J. Schmahl, R.A. Schwartz, Measurement and interpretation of X-ray visibilities with RHESSI, in *AGU Spring Meeting Abstracts*, vol. 2005 (2005), pp. SP21A–12
98. G.J. Hurford, S. Krucker, R.P. Lin, R.A. Schwartz, G.H. Share, D.M. Smith, Gamma-Ray imaging of the 2003 October/November solar flares. Astrophys. J. Lett. **644**, L93–L96 (2006)
99. E.T. Jaynes, Information theory and statistical mechanics. Phys. Rev. **106**(4), 620–630 (1957)
100. N.L.S. Jeffrey, The Spatial, Spectral and Polarization Properties of Solar Flare X-ray Sources. PhD Thesis, University of Glasgow, 2014
101. N.L.S. Jeffrey, P. Saint-Hilaire, E.P. Kontar, Probing solar flare accelerated electron distributions with prospective X-ray polarimetry missions. Astron. Astrophys. **642**, A79 (2020)
102. J. Kašparová, E.P. Kontar, J.C. Brown, Hard X-ray spectra and positions of solar flares observed by rhessi: photospheric albedo, directivity and electron spectra. Astron. Astrophys. **466**, 705–712 (2007)
103. L. Kipp, M. Skibowski, R.L. Johnson, R. Berndt, R. Adelung, S. Harm, R. Seemann, Sharper images by focusing soft X-rays with photon sieves. Nature **414**(6860), 184–188 (2001)
104. K. Kobayashi, J. Cirtain, A.R. Winebarger, K. Korreck, L. Golub, R.W. Walsh, B. De Pontieu, C. DeForest, A. Title, S. Kuzin, S. Savage, D. Beabout, B. Beabout, W. Podgorski, D. Caldwell, K. McCracken, M. Ordway, H. Bergner, R. Gates, S. McKillop, P. Cheimets, S. Platt, N. Mitchell, D. Windt, The high-resolution coronal imager (Hi-C). Solar Phys. **289**(11), 4393–4412 (2014)
105. H.W. Koch, J.W. Motz, Bremsstrahlung cross-section formulas and related data. Rev. Modern Phys. **31**, 920–955 (1959)
106. E.P. Kontar, N.L.S. Jeffrey, Positions and sizes of X-ray solar flare sources. Astron. Astrophys. **513**, L2+ (2010)
107. E.P. Kontar, M. Piana, A.M. Massone, A.G. Emslie, J.C. Brown, Generalized regularization techniques with constraints for the analysis of solar bremsstrahlung X-ray spectra. Solar Phys. **225**(2), 293–309 (2004)
108. E.P. Kontar, A.G. Emslie, M. Piana, A.M. Massone, J.C. Brown, Determination of electron flux spectra in a solar flare with an augmented regularization method: application to RHESSI data. Solar Phys. **226**(2), 317–325 (2005)
109. E.P. Kontar, A.G. Emslie, A.M. Massone, M. Piana, J.C. Brown, M. Prato, Electron-electron bremsstrahlung emission and the inference of electron flux spectra in solar flares. Astrophys. J. **670**, 857–861 (2007)
110. E.P. Kontar, I.G. Hannah, N.L.S. Jeffrey, M. Battaglia, The sub-arcsecond hard X-ray structure of loop footpoints in a solar flare. Astrophys. J. **717**, 250–256 (2010)
111. E.P. Kontar, J.C. Brown, A.G. Emslie, W. Hajdas, G.D. Holman, G.J. Hurford, J. Kašparová, P.C.V. Mallik, A.M. Massone, M.L. McConnell, M. Piana, M. Prato, E.J. Schmahl, E. Suarez-Garcia, Deducing electron properties from hard X-ray observations. Space Sci. Rev. **159**, 301–355 (2011)
112. R.A. Kopp, G.W. Pneuman, Magnetic reconnection in the corona and the loop prominence phenomenon. Solar Phys. **50**, 85–98 (1976)
113. T. Kosugi, K. Makishima, T. Murakami, T. Sakao, T. Dotani, M. Inda, K. Kai, S. Masuda, H. Nakajima, Y. Ogawara, M. Sawa, K. Shibasaki, The hard X-ray telescope (HXT) for the SOLAR-A mission. Solar Phys. **136**(1), 17–36 (1991)
114. S. Krucker, R.P. Lin, Relative timing and spectra of solar flare hard X-ray sources. Solar Phys. **210**(1), 229–243 (2002)

115. S. Krucker, G.J. Hurford, R.P. Lin, Hard X-Ray source motions in the 2002 July 23 Gamma-Ray Flare. Astrophys. J. Lett. **595**(2), L103–L106 (2003)
116. S. Krucker, M. Battaglia, P.J. Cargill, L. Fletcher, H.S. Hudson, A.L. MacKinnon, S. Masuda, L. Sui, M. Tomczak, A.L. Veronig, L. Vlahos, S.M. White, Hard X-ray emission from the solar corona. Astron. Astrophys. Rev. **16**, 155–208 (2008)
117. S. Krucker, G.J. Hurford, A.L. MacKinnon, A.Y. Shih, R.P. Lin, Coronal γ-ray bremsstrahlung from solar flare-accelerated electrons. Astrophys. J. Lett. **678**, L63–L66 (2008)
118. S. Krucker, E.P. Kontar, S. Christe, L. Glesener, R.P. Lin, Electron acceleration associated with solar jets. Astrophys. J. **742**(2), 82 (2011)
119. S. Krucker, S. Christe, L. Glesener, S. Ishikawa, S. McBride, D. Glaser, P. Turin, R.P. Lin, M. Gubarev, B. Ramsey, S. Saito, Y. Tanaka, T. Takahashi, S. Watanabe, T. Tanaka, H. Tajima, S. Masuda, The focusing optics X-ray solar imager (FOXSI), in *Society of Photo-Optical Instrumentation Engineers (SPIE) Conference Series*, ed. by S.L. O'Dell, G. Pareschi, vol. 8147 (2011), p. 814705
120. S. Krucker, G.J. Hurford, O. Grimm, S. Kögl, H.P. Gröbelbauer, L. Etesi, D. Casadei, A. Csillaghy, A.O. Benz, N.G. Arnold, F. Molendini, P. Orleanski, D. Schori, H. Xiao, M. Kuhar, N. Hochmuth, S. Felix, F. Schramka, S. Marcin, S. Kobler, L. Iseli, M. Dreier, H.J. Wiehl, L. Kleint, M. Battaglia, E. Lastufka, H. Sathiapal, K. Lapadula, M. Bednarzik, G. Birrer, St. Stutz, Ch. Wild, F. Marone, K.R. Skup, A. Cichocki, K. Ber, K. Rutkowski, W. Bujwan, G. Juchnikowski, M. Winkler, M. Darmetko, M. Michalska, K. Seweryn, A. Białek, P. Osica, J. Sylwester, M. Kowalinski, D. Ścisłowski, M. Siarkowski, M. Stęślicki, T. Mrozek, P. Podgórski, A. Meuris, O. Limousin, O. Gevin, I. Le Mer, S. Brun, A. Strugarek, N. Vilmer, S. Musset, M. Maksimović, F. Fárník, Z. Kozáček, J. Kašparová, G. Mann, H. Önel, A. Warmuth, J. Rendtel, J. Anderson, S. Bauer, F. Dionies, J. Paschke, D. Plüschke, M. Woche, F. Schuller, A.M. Veronig, E.C.M. Dickson, P.T. Gallagher, S.A. Maloney, D.S. Bloomfield, M. Piana, A.M. Massone, F. Benvenuto, P. Massa, R.A. Schwartz, B.R. Dennis, H.F. van Beek, J. Rodríguez-Pacheco, R.P. Lin, The spectrometer/telescope for imaging X-rays (STIX). Astron. Astrophys. **642**, A15 (2020)
121. T.N. Larosa, A.G. Emslie, Beam-return current systems in nonthermal solar flare models. Solar Phys. **120**, 343–349 (1989)
122. E. Lastufka, D. Casadei, G. Hurford, M. Kuhar, G. Torre, S. Krucker, The micro solar flare apparatus (MiSolFA) instrument concept. Adv. Space Res. **66**(1), 10–20 (2020)
123. J. Leach, V. Petrosian, A.G. Emslie, The interpretation of hard X-ray polarization measurements in solar flares. Solar Phys. **96**, 331–337 (1985)
124. J.R. Lemen, A.M. Title, D.J. Akin, P.F. Boerner, C. Chou, J.F. Drake, D.W. Duncan, C.G. Edwards, F.M. Friedlaender, G.F. Heyman, N.E. Hurlburt, N.L. Katz, G.D. Kushner, M. Levay, R.W. Lindgren, D.P. Mathur, E.L. McFeaters, S. Mitchell, R.A. Rehse, C.J. Schrijver, L.A. Springer, R.A. Stern, T.D. Tarbell, J.-P. Wuelser, C.J. Wolfson, C. Yanari, J.A. Bookbinder, P.N. Cheimets, D. Caldwell, E.E. Deluca, R. Gates, L. Golub, S. Park, W.A. Podgorski, R.I. Bush, P.H. Scherrer, M.A. Gummin, P. Smith, G. Auker, P. Jerram, P. Pool, R. Soufli, D.L. Windt, S. Beardsley, M. Clapp, J. Lang, N. Waltham, The atmospheric imaging assembly (AIA) on the solar dynamics observatory (SDO). Solar Phys. **275**, 17–40 (2012)
125. R.P. Lin, B.R. Dennis, G.J. Hurford, D.M. Smith, A. Zehnder, P.R. Harvey, D.W. Curtis, D. Pankow, P. Turin, M. Bester, A. Csillaghy, M. Lewis, N. Madden, H.F. van Beek, M. Appleby, T. Raudorf, J. McTiernan, R. Ramaty, E. Schmahl, R. Schwartz, S. Krucker, R. Abiad, T. Quinn, P. Berg, M. Hashii, R. Sterling, R. Jackson, R. Pratt, R.D. Campbell, D. Malone, D. Landis, C.P. Barrington-Leigh, S. Slassi-Sennou, C. Cork, D. Clark, D. Amato, L. Orwig, R. Boyle, I.S. Banks, K. Shirey, A.K. Tolbert, D. Zarro, F. Snow, K. Thomsen, R. Henneck, A. McHedlishvili, P. Ming, M. Fivian, John Jordan, Richard Wanner, Jerry Crubb, J. Preble, M. Matranga, A. Benz, H. Hudson, R.C. Canfield, G.D. Holman, C. Crannell, T. Kosugi, A.G. Emslie, N. Vilmer, J.C. Brown, C. Johns-Krull, M. Aschwanden, T. Metcalf, A. Conway, The Reuven Ramaty high-energy solar spectroscopic imager (RHESSI). Solar Phys. **210**(1), 3–32 (2002)

126. R.P. Lin, A. Caspi, S. Krucker, H. Hudson, G. Hurford, S. Bandler, S.D. Christe, J. Davila, B. Dennis, G. Holman, R. Milligan, A.Y. Shih, S. Kahler, E. Kontar, M. Wiedenbeck, J. Cirtain, G. Doschek, G.H. Share, A. Vourlidas, J. Raymond, D.M. Smith, M. McConnell, A.G. Emslie, Solar Eruptive Events (SEE) 2020 Mission Concept (2013). arXiv e-prints, page arXiv:1311.5243
127. M.E. Machado, E.H. Avrett, J.E. Vernazza, R.W. Noyes, Semiempirical models of chromospheric flare regions. Astrophys. J. **242**, 336–351 (1980)
128. J.T. Mariska, A.G. Emslie, P. Li, Numerical simulations of impulsively heated solar flares. Astrophys. J. **341**, 1067–1074 (1989)
129. D. Martínez-Galarce, R. Soufli, D.L. Windt, M. Bruner, E. Gullikson, S. Khatri, E. Spiller, J.C. Robinson, S. Baker, E. Prast, Multisegmented, multilayer-coated mirrors for the solar ultraviolet imager. Optical Eng. **52**, 095102 (2013)
130. P. Massa, M. Piana, A.M. Massone, F. Benvenuto, Count-based imaging model for the spectrometer/telescope for imaging X-rays (STIX) in solar orbiter. Astron. Astrophys. **624**, A130 (2019)
131. P. Massa, R. Schwartz, A.K. Tolbert, A.M. Massone, B.R. Dennis, M. Piana, F. Benvenuto, MEM_GE: a new maximum entropy method for image reconstruction from solar X-ray visibilities. Astrophys. J. **894**(1), 46 (2020)
132. A.M. Massone, A.G. Emslie, E.P. Kontar, M. Piana, M. Prato, J.C. Brown, Anisotropic bremsstrahlung emission and the form of regularized electron flux spectra in solar flares. Astrophys. J. **613**, 1233–1240 (2004)
133. A.M. Massone, A.G. Emslie, G.J. Hurford, M. Prato, E.P. Kontar, M. Piana, Hard X-ray imaging of solar flares using interpolated visibilities. Astrophys. J. **703**, 2004–2016 (2009)
134. S. Masuda, T. Kosugi, H. Hara, S. Tsuneta, Y. Ogawara, A loop-top hard X-ray source in a compact solar flare as evidence for magnetic reconnection. Nature **371**, 495–497 (1994)
135. M.L. McConnell, J.M. Ryan, D.M. Smith, R.P. Lin, A.G. Emslie, RHESSI as a hard X-ray polarimeter. Solar Phys. **210**, 125–142 (2002)
136. M.L. McConnell, D.M. Smith, A.G. Emslie, G.J. Hurford, R.P. Lin, J.M. Ryan, Hard X-ray solar flare polarimetry with RHESSI. Adv. Space Res. **34**, 462–466 (2004)
137. D.B. Melrose, *Plasma Astrophysics: Nonthermal Processes in Diffuse Magnetized Plasmas. Volume 1 - The Emission, Absorption and Transfer of Waves in Plasmas* (Gordon and Breach, London, 1980)
138. D.B. Melrose, *Plasma Astrophysics: Nonthermal Processes in Diffuse Magnetized Plasmas. Volume 2 - Astrophysical Applications* (Gordon and Breach, London, 1980)
139. T.R. Metcalf, H.S. Hudson, T. Kosugi, R.C. Puetter, R.K. Pina, Pixon-based multiresolution image reconstruction for Yohkoh's hard X-ray telescope. Astrophys. J. **466**, 585 (1996)
140. K. Miller, Least squares methods for Ill-posed problems with a prescribed bound. SIAM J. Math. Anal. **1**, 52 (1970)
141. N. Narakage, M. Oka, Y. Fukazawa, K. Matsuzaki, S. Watanabe, T. Sakao, K. Hagino, I. Mitsuishi, T. Mizuno, I. Shinohara, M. Shimojo, S. Takasao, H. Tanabe, M. Ueno, T. Takahashi, T. Takashima, M Ohta, Satellite mission: PhoENiX (Physics of Energetic and Non-thermal plasmas in the X (= magnetic reconnection) region), in *Space Telescopes and Instrumentation 2020: Ultraviolet to Gamma Ray*, ed. by J.-W.A. den Herder, S. Nikzad, K. Nakazawa, vol. 11444 (International Society for Optics and Photonics, SPIE, Bellingham, 2020)
142. N. Narukage, Satellite mission: PhoENiX (Physics of Energetic and Non-thermal plasmas in the X (= magnetic reconnection) region), in *American Astronomical Society Meeting Abstracts #234*, vol. 234 (2019), p. 126.03
143. F. Natterer, G. Wang, The mathematics of computerized tomography. Med. Phys. **29**(1), 107–108 (2002)
144. F. Natterer, F. Wübbeling, *Mathematical Methods in Image Reconstruction.* (Society for Industrial and Applied Mathematics, Philadelphia, 2001)
145. M. Oda, High-resolution X-ray collimator with broad field of view for astronomical use. Appl. Optics **4**, 143–143 (1965)

146. M. Oda, X-ray imaging techniques - modulation collimator and coded mask. Adv. Space Res. **2**, 207–216 (1983)
147. H. Odaka, T. Kasuga, K. Hatauchi, T. Tamba, S. Takashima, H. Suzuki, Y. Aizawa, A. Bamba, T. Watanabe, S. Nammoku, A. Tanimoto, Concept of a CubeSat-based hard X-ray imaging polarimeter: Cipher, in *Society of Photo-Optical Instrumentation Engineers (SPIE) Conference Series*, vol. 11444 (2020), p. 114445V
148. B.T. Park, V. Petrosian, R.A. Schwartz, Stochastic acceleration and photon emission in electron-dominated solar flares. Astrophys. J. **489**(1), 358–366 (1997)
149. E. Perracchione, A.M. Massone, M. Piana, Feature augmentation for the inversion of the Fourier transform with limited data. Inverse Problems **37**(10), 105001 (2021)
150. M. Piana, Inversion of bremsstrahlung spectra emitted by solar plasma. Astron. Astrophys. **288**, 949–959 (1994)
151. M. Piana, M. Bertero, Regularized deconvolution of multiple images of the same object. J. Opt. Soc. Amer. A **13**(7), 1516–1523 (1996)
152. M. Piana, M. Bertero, Projected landweber method and preconditioning. Inverse Problems **13**(2), 441–463 (1997)
153. M. Piana, A.M. Massone, E.P. Kontar, A.G. Emslie, J.C. Brown, R.A. Schwartz, Regularized electron flux spectra in the 2002 July 23 solar flare. Astrophys. J. Lett. **595**, L127–L130 (2003)
154. M. Piana, A.M. Massone, G.J. Hurford, M. Prato, A.G. Emslie, E.P. Kontar, R.A. Schwartz, Electron flux spectral imaging of solar flares through regularized analysis of hard X-ray source visibilities. Astrophys. J. **665**, 846–855 (2007)
155. R.K. Pina, R.C. Puetter, Bayesian image reconstruction - the pixon and optimal image modeling. Publ. Astron. Soc. Pacif. **105**, 630–637 (1993)
156. M. Prato, M. Piana, J.C. Brown, A.G. Emslie, E.P. Kontar, A.M. Massone, Regularized reconstruction of the differential emission measure from solar flare hard X-ray spectra. Solar Phys. **237**(1), 61–83 (2006)
157. M. Prato, M. Piana, A.G. Emslie, G.J. Hurford, E.P. Kontar, A.M. Massone, A regularized visibility-based approach to astronomical imaging spectroscopy. SIAM J. Imag. Sci. **2**(3), 910–930 (2009)
158. E. Priest, T. Forbes, *Magnetic Reconnection* (Cambridge University Press, Cambridge, 2000)
159. T.A. Prince, G.J. Hurford, H.S. Hudson, C.J. Crannell. Gamma-Ray and hard X-ray imaging of solar flares. Solar Phys. **118**(1–2), 269–290 (1988)
160. R.C. Puetter, Pixon-based multiresolution image reconstruction and the quantification of picture information content. Int. J. Imag. Syst. Technol. **6**(4), 314–331 (1995)
161. D.F. Ryan, P.C. Chamberlin, R.O. Milligan, P.T. Gallagher, Decay-phase cooling and inferred heating of M- and X-class solar flares. Astrophys. J. **778**, 68 (2013)
162. P. Saint-Hilaire, S. Krucker, R.P. Lin, A statistical survey of hard X-ray spectral characteristics of solar flares with two footpoints. Solar Phys. **250**, 53–73 (2008)
163. P. Saint-Hilaire, N.L.S. Jeffrey, J.C. Martinez Oliveros, A.Y. Shih, A. Zoglauer, A. Caspi, P. Lichtmacher, S. Boggs, G.J. Hurford, S. Krucker, J.G. Sample, A. Tremsin, Solar flare hard X-ray and Gamma-ray imaging spectro-polarimetry with GRIPS and SAPPHIRE/SHARPIE, in *AGU Fall Meeting Abstracts*, vol. 2019 (2019), pp. SH31C–3310
164. J. Sato, T. Kosugi, K. Makishima, Improvement of YOHKOH hard X-ray imaging. Publ. Astron. Soc. Jpn **51**, 127–150 (1999)
165. S. Savage, A. Winebarger, L. Glesener, L. Golub, P. Chamberlin, Hi-C Flare Team, FOXSI-4 Team, Snifs Team, The first solar flare sounding rocket campaign and its potential impacts for high energy solar instrumentation, in *American Astronomical Society Meeting Abstracts*, vol. 53 (2021), p. 313.15
166. V. Schoenfelder, H. Aarts, K. Bennett, H. de Boer, J. Clear, W. Collmar, A. Connors, A. Deerenberg, R. Diehl, A. von Dordrecht, J.W. den Herder, W. Hermsen, M. Kippen, L. Kuiper, G. Lichti, J. Lockwood, J. Macri, M. McConnell, D. Morris, R. Much, J. Ryan, G. Simpson, M. Snelling, G. Stacy, H. Steinle, A. Strong, B.N. Swanenburg, B. Taylor, C. de Vries, C. Winkler, Instrument description and performance of the imaging Gamma-Ray

telescope COMPTEL aboard the compton Gamma-Ray observatory. Astrophys. J. Suppl. **86**, 657 (1993)
167. R.A. Schwartz, The effect of pulse-pileup on RHESSI image reconstruction, in *AGU Spring Meeting Abstracts*, vol. 2008 (2008), pp. SP51B–03
168. R.A. Schwartz, A. Csillaghy, A.K. Tolbert, G.J. Hurford, J. McTiernan, D. Zarro, RHESSI data analysis software: rationale and methods. Solar Phys. **210**(1), 165–191 (2002)
169. F. Sciacchitano, A. Sorrentino, A.G. Emslie, A.M. Massone, M. Piana, Identification of multiple hard X-ray sources in solar flares: A Bayesian analysis of the 2002 February 20 event. Astrophys. J. **862**(1), 68 (2018)
170. L.A. Shepp, Y.Vardi, Maximum likelihood reconstruction for emission tomography. IEEE Trans. Med. Imag. **1**(2), 113–122 (1982)
171. G.A. Shields, A brief history of active galactic nuclei. Publ. Astron. Soc. Pac. **111**(760), 661–678 (1999)
172. A.Y. Shih, L. Glesener, S. Christe, K. Reeves, S. Gburek, M. Alaoui, J.C. Allred, W. Baumgartner, A. Caspi, B.R. Dennis, J.F. Drake, K. Goetz, L. Golub, S.E. Guidoni, A. Inglis, I.G. Hannah, G. Holman, L. Hayes, J. Ireland, G.S. Kerr, J.A. Klimchuk, S. Krucker, D.E. McKenzie, C.S. Moore, S. Musset, J.W. Reep, D. Ryan, P. Saint-Hilaire, S.L. Savage, D.B. Seaton, M. Steslicki, T.N. Woods, Combined next-generation X-ray and EUV observations with the FIERCE mission concept, in *AGU Fall Meeting Abstracts*, vol. 2019 (2019), pp. SH33A–08
173. S. Singer, J.Nelder, Nelder-Mead algorithm. Scholarpedia **4**(7), 2928 (2009)
174. G.K. Skinner, Diffractive/refractive optics for high energy astronomy. I. Gamma-ray Phase fresnel lenses. Astron. Astrophys. **375**, 691–700 (2001)
175. G.K. Skinner, Diffractive X-ray telescopes. *X-Ray Optics and Instrumentation, 2010. Special Issue on X-Ray Focusing: Techniques and Applications, id.743485* (2010)
176. G.K. Skinner, B.R. Dennis, J.F. Krizmanic, E.P. Kontar, Science enabled by high precision inertial formation flying. Int. J. Space Sci. Eng. **1**, 331 (2013)
177. C.C. Sleator, A. Zoglauer, A.W. Lowell, C.A. Kierans, N. Pellegrini, J. Beechert, S.E. Boggs, T.J. Brandt, H. Lazar, J.M. Roberts, T. Siegert, J.A. Tomsick, Benchmarking simulations of the compton spectrometer and imager with calibrations. Nuclear Instrum. Methods Phys. Res. A **946**, 162643 (2019)
178. D.M. Smith, R.P. Lin, P. Turin, et al. The RHESSI spectrometer. Solar Phys. **210**, 33–60 (2002)
179. D.M. Smith, G.H. Share, R.J. Murphy, R.A. Schwartz, A.Y. Shih, R.P. Lin, High-resolution spectroscopy of Gamma-Ray lines from the X-class solar flare of 2002 July 23. Astrophys. J. Lett. **595**, L81–L84 (2003)
180. L. Spitzer, *Physics of Fully Ionized Gases* (Interscience, New York, 1962)
181. A. Strong, W. Collmar, COMPTEL reloaded: a heritage project in MeV astronomy. Mem. Soc. Astron. Italiana **90**, 297 (2019)
182. P.A. Sturrock, A model of solar flares, in *Structure and Development of Solar Active Regions*, ed.by K.O. Kiepenheuer, vol. 35. IAU Symposium (1968), p. 471
183. J. Sylwester, I. Gaicki, Z. Kordylewski, M. Nowak, S. Kowalinski, M. Sjarkowski, W. Bentley, R.D. Trzebinski, M.W. Whyndham, P.R. Guttridge, J.L. Culhane, J. Lang, K.J.H. Phillips, C.M. Brown, G.A. Doschek, V.N. Oraevsky, S.I. Boldyrev, I.M. Kopaev, A.I. Stepanov, V.Yu. Klepikov, RESIK: high sensitivity soft X-ray spectrometer for the study of solar flare plasma, in *Crossroads for European Solar and Heliospheric Physics. Recent Achievements and Future Mission Possibilities*, ESA Special Publication, vol. 417 (1998), p. 313
184. T. Takakura, K. Ohki, S. Tsuneta, N. Nitta, Hard X-ray images of impulsive bursts. Solar Phys. **86**(1–2), 323–331 (1983)
185. E. Tandberg-Hanssen, A.G. Emslie, *The Physics of Solar Flares* (Cambridge University Press, Cambridge, 1988)
186. A.R. Thompson, J.M. Moran, G.W. Swenson, Jr, *Interferometry and Synthesis in Radio Astronomy*, , 2nd edn. (Wiley, Hoboken, 2001)

187. A.N. Tikhonov, On the solution of incorrectly formulated problems and the regularization method. Soviet Math. Dokl. **4**, 1035–1038 (1963)
188. A.N. Tikhonov, A.V. Goncharsky, V.V. Stepanov, A.G. Yagola, *Numerical Methods for the Solution of Ill-posed Problems*, , vol. 328 (Springer Science & Business Media, Berlin, 2013)
189. R.C. Tolman, *The Principles of Statistical Mechanics*. (Clarendon Press, Oxford, 1938)
190. V.M. Tomozov, Plasma processes in solar flares, in *Basic Plasma Processes on the Sun, IAU Symposium*, ed. by E.R. Priest, V. Krishan (1990), p. 355
191. G. Torre, N. Pinamonti, A.G. Emslie, J. Guo, A.M. Massone, M. Piana, Empirical determination of the energy loss rate of accelerated electrons in a well-observed solar flare. Astrophys. J. **751**(2), 129 (2012)
192. K. Van Audenhaege, R. Van Holen, S. Vandenberghe, C. Vanhove, S.D. Metzler, S.C. Moore, Review of SPECT collimator selection, optimization, and fabrication for clinical and preclinical imaging. Med. Phys. **42**(8), 4796–4813 (2015)
193. H.F. van Beek, P. Hoyng, B. Lafleur, G.M. Simnett, The hard X-ray imaging spectrometer (HXIS). Solar Phys. **65**(1), 39–52 (1980)
194. H.F. van Beek, C. de Jager, A. Schadee, Z. Svestka, A. Boelee, A. Duijveman, M. Galama, R. Hoekstra, P. Hoyng, R. Fryer, G.M. Simnett, J.P. Imhof, H. LaFleur, H.V.A.M. Maseland, W.M. Mels, J. Schrijver, J.J.M. van der Laan, P. van Rens, W. van Tend, F. Werkhoven, A.P. Willmore, J.W.G. Wilson, M.E. Machado, W. Zandee, The Limb Flare of 1980 April 30 as seen by the hard X-ray imaging spectrometer. Astrophys. J. Lett. **244**, L157–L162 (1981)
195. G.H.J. van den Oord, The electrodynamics of beam/return current systems in the solar corona. Astron. Astrophys. **234**, 496–518 (1990)
196. J.E. Vernazza, E.H. Avrett, R. Loeser, Structure of the solar chromosphere. III. Models of the EUV brightness components of the quiet sun. Astrophys. J. Suppl. **45**, 635–725 (1981)
197. J.T. Vievering, L. Glesener, P.S. Athiray, J.C. Buitrago-Casas, S. Musset, D.F. Ryan, S. Ishikawa, J. Duncan, S. Christe, S. Krucker, FOXSI-2 solar microflares. II. Hard X-ray imaging spectroscopy and flare energetics. Astrophys. J. **913**(1), 15 (2021)
198. A. Warmuth, G. Mann, Thermal and nonthermal hard X-ray source sizes in solar flares obtained from RHESSI observations. I. Observations and evaluation of methods. Astron. Astrophys. **552**, A86 (2013)
199. A. Warmuth, G. Mann, Thermal-nonthermal energy partition in solar flares derived from X-ray, EUV, and bolometric observations. discussion of recent studies. Astron. Astrophys. **644**, A172 (2020)
200. J.G. Webster (ed.), *Medical Instrumentation Application and Design*, 4th edn. (Wiley, Hoboken, 2010)
201. M.C. Weisskopf, B. Ramsey, S. O'Dell, A. Tennant, R. Elsner, P. Soffitta, R. Bellazzini, E. Costa, J. Kolodziejczak, V. Kaspi, F. Muleri, H. Marshall, G. Matt, R. Romani, The imaging X-ray polarimetry explorer (IXPE), in *Space Telescopes and Instrumentation 2016: Ultraviolet to Gamma Ray*, ed. by J.-W.A. den Herder, T. Takahashi, M. Bautz. Society of Photo-Optical Instrumentation Engineers (SPIE) Conference Series, vol. 9905 (2016), p. 990517
202. D.L. Windt, Advancements in hard X-ray multilayers for X-ray astronomy, in *Society of Photo-Optical Instrumentation Engineers (SPIE) Conference Series*, vol. 9603 (2015), p. 96031C
203. H. Wolter, Spiegelsysteme streifenden Einfalls als abbildende Optiken für Röntgenstrahlen. Annalen der Physik **445**(1–2), 94–114 (1952)
204. H. Wolter, Verallgemeinerte Schwarzschildsche Spiegelsysteme streifender Reflexion als Optiken für Röntgenstrahlen. Annalen der Physik **445**(4–5), 286–295 (1952)
205. H. Wolter, Mirror systems with grazing incidence as image-forming optics for X-rays. (Translated into English from Ann. Phys. (Leipzig), ser. 6, **10**, 94–114 (1952)) **10**, 94–114 (1975)
206. Y. Xu, A.G. Emslie, G.J. Hurford, RHESSI hard X-ray imaging spectroscopy of extended sources and the physical properties of electron acceleration regions in solar flares. Astrophys. J. **673**, 576–585 (2008)

207. Z. Zhang, J. Wu, HXI onboard ASO-S: mapping solar flares in hard X-ray band, in *42nd COSPAR Scientific Assembly*, vol. 42 (2018), pp. D2.1–2–18
208. Z. Zhang, D.-Y. Chen, J. Wu, J. Chang, Y.-M. Hu, Y. Su, Y. Zhang, J.-P. Wang, Y.-M. Liang, T. Ma, J.-H. Guo, M.-S. Cai, Y.-Q. Zhang, Y.-Y. Huang, X.-Y. Peng, Z.-B. Tang, X. Zhao, H.-H. Zhou, L.-G. Wang, J.-X. Song, M. Ma, G.-Z. Xu, J.-F. Yang, D. Lu, Y.-H. He, J.-Y. Tao, X.-L. Ma, B.-G. Lv, Y.-P. Bai, C.-X. Cao, Y. Huang, W.-Q. Gan, Hard X-ray imager (HXI) onboard the ASO-S mission. Res. Astron. Astrophys. **19**(11), 160 (2019)
209. P. Zhang, W. Hajdas, S.-M. Liu, Y. Su, Y.-P. Li, W. Chen. In-flight low energy X-ray calibration of POLAR detector on TianGong2. Chinese Astron. Astrophys. **44**(1), 87–104 (2020)
210. V.V. Zharkova, M. Gordovskyy, The kinetic effects of electron beam precipitation and resulting hard X-ray intensity in solar flares. Astron. Astrophys. **432**(3), 1033–1047 (2005)
211. V.V. Zharkova, M. Gordovskyy, The effect of the electric field induced by precipitating electron beams on hard X-ray photon and mean electron spectra. Astrophys. J. **651**, 553–565 (2006)
212. V.V. Zharkova, K. Arzner, A.O. Benz, P. Browning, C. Dauphin, A.G. Emslie, L. Fletcher, E.P. Kontar, G. Mann, M. Onofri, V. Petrosian, R. Turkmani, N. Vilmer, L. Vlahos, Recent advances in understanding particle acceleration processes in solar flares. Space Sci. Rev. **159**, 357–420 (2011)

Index

M. Piana et al., *Hard X-Ray Imaging of Solar Flares*,
https://doi.org/10.1007/978-3-030-87277-9

Printed in the United States
by Baker & Taylor Publisher Services